Secret Agents Future

Going Where There Be Dragons

Ken Stange

BOOKS BY

KEN STANGE

A Smoother Pebble, A Prettier Shell (Penumbra Press)

Advice To Travellers (Penumbra Press)

Bourgeois Pleasures (Quarry Press)

Bushed (York Publishing)

Cold Pigging Poetics (York Publishing)

Colonization Of a Cold Planet (Two Cultures Press)

Embracing The Moon: 25 Little Worlds (Two Cultures Press)

Explaining Canada: A Primer For Yanks (Two Cultures Press)

God When He's Drunk (Two Cultures Press)

Going Home (Two Cultures Press)

Love Is A Grave (Nebula Press)

More Than Ample (Two Cultures Press)

Nocturnal Rhythms (Penumbra Press)

The Sad Science Of Love (Two Cultures Press)

*Secret Agents Past: The Parting Of The Waters (*Two Cultures Press)

*Secret Agents Present: Looking Through A Glass Darkly (*Two Cultures Press)

These Proses A Problem Or Two (Two Cultures Press)

Secret Agents Future

Going Where There Be Dragons

~~

Ken Stange

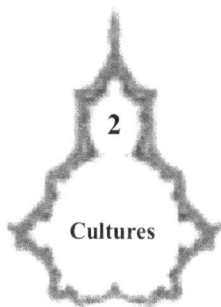

Two Cultures Press

2014

For information about permission to reprint, record, or perform sections of this book, write to **Two Cultures Press**, 970 Copeland, North Bay, Ontario, Canada, P1B 3E4 (**info@twoculturespress.com**)

Library and Archives Canada Cataloguing in Publication

Stange, Ken, 1946-, author
        Secret agents future : going where there be dragons / Ken Stange. -- 1st edition.

Includes bibliographical references.
ISBN 978-0-9939201-2-7 (pbk.)

        1. Creative ability. 2. Creative ability in science. 3. Art and science. I. Title.

BF408.S725 2014   153.3'5   C2014-906736-4

Acknowledgements

"Redefining Creativity: How Science and Technology Make Us Rethink Creativity" (based on some ideas in this book) was presented and broadcast as a TEDx lecture sponsored by Nipissing University. (2012)

"Math, Art and Science: The Two Cultures Coming Together" (the basis for some of the ideas in this book) was presented at the OAME (Ontario Association for Mathematics Education) 25th Annual Conference in North Bay, Ontario. (1998)

"Intertwining The Two Cultures In The Year Two Thousand", (based on ideas in this book) was used in a paper presented at The Second International Conference Mathematics & Design Conference in Bilbao, Spain. (1998)

Cover Art and Design: Ken Stange

ISBN: 978-0-9939201-2-7

This book is dedicated to my not-so-secret agent: Ursula. Her collaboration in all my creative efforts is a secret all over the block.

Contents

FOREWORD: FOREWARNING

This book is the last in a trilogy on the nature of creativity in the arts and in the sciences. Because organizing my many scattered thoughts on this topic wasn't easy, I eventually decided to adopt a quasi-chronological approach. The first two books were about the past and the present. This book consists of speculations about changes that might affect future creative endeavours.

The three *Secret Agents* books:
>*Secret Agents Past: The Parting Of The Waters*
>*Secret Agents Present: Looking Through A Glass Darkly*
>*Secret Agents Future: Going Where There Be Dragons*

PREFACE: CAVEAT LECTOR

This book is intended, as were the previous two, for the general reader and not as a scholarly review of 'the literature' on creativity. It is intended as an informal, speculative quasi-philosophical exploration of the nature of creativity in art and in science, spiced up with a few polemics. For this reason, and for the sake of readability, I have kept the customary, often cumbersome, paraphernalia of formal scholarship to a minimum; so the reader will not find any of those intrusive APA format parentheses with the names and dates of research papers. Nor have I banned the use of the first person singular and personal anecdote from my prose, as is required in scientific publications trying to maintain an objective tone. I openly admit to being opinionated and less than coolly objective, for I care passionately about the topic of this book. For the same reason, I've tried to avoid the "Hey, I'm not responsible" passive voice.

Consequently the reader won't find extensive, formally constructed references to support all of the allegedly factual information included in this book. However, most of what I say, not opine, can be easily confirmed independently, given access to The Internet and any decent library. In those cases where some fact or reference is not so easily confirmed by the average reader in our wonderful electronic age, or where I felt it worthwhile to point to interesting material relevant to the topic at hand, I have inserted an adequate bibliographical pointer in a footnote. In most cases I am sure that simply giving the author and the title of a reference these days is sufficient to easily locate full bibliographical information on the Internet.

Thus the policy I have adopted is to use footnotes primarily as a place for extended parenthetical information or comment, the occasional wisecrack that I couldn't resist but didn't want to interrupt the flow of the main text, and brief author/title pointers to relevant books or articles. Only when directly quoting a writer or when referring to extremely specific material—such as a particular research study—have I felt it necessary to include a detailed bibliographical reference in the footnote.

Naturally, I accept full responsibility for any errors of fact or interpretation of fact. That I've not called expert witnesses to substantiate all my statements should not be interpreted as my not checking my information. Those who wish to question something I've passed off as factual should find it easy to check up on me, and I'm more than willing to stand corrected. I'm sure there are many experts in the various fields where I've trespassed that will queue up to do just that—and shoot me down. I only ask they don't shoot to kill.

.

So although this book has no references page, I couldn't resist appending recommended further reading: I have included a somewhat eccentric, quite eclectic, selected bibliography, sorted by chapter topic. Many of these works have been a source of both factual information and inspiration to speculation.

PROLOGUE: THE DOOR TO THE FUTURE

"The future belongs to those who believe in the beauty of their dreams."
—*Eleanor Roosevelt*

"I feel very adventurous. There are so many doors to be opened, and I'm not afraid to look behind them."
—*Elizabeth Taylor*

The future is an unopened door. There may be fire-breathing dragons waiting, but they'll only be guarding the portal to a new world.
.

Creativity involves opening that door, and, at least till now, it has lead us to better place. However, the need to explore has to be greater than fear of the dangers and risks involved. Certainly the risks are real and many have been burned.
.

People who are creative may not be braver; they might just be more foolish. Ignoring risk is not necessarily admirably bold. However, it is unavoidable for accomplishing many things. (Certainly for some risk is innately appealing, but they are few.) We all take risks, because living life to its fullest is a risk. We choose our risks by the satisfaction it gives us, and our choices are not necessarily wise. Only for some is taking risks inherently satisfying; but for most it is just worth the satisfaction that requires it.
.

For what it's worth, psychometric tests seem to suggest that the creative are less risk aversive than average. It is a prerequisite to exploring new territory. Columbus was not dissuaded from setting sail because of a fear of the warnings inscribed on early maps for uncharted areas: "Here Be Dragons".*
.

---

* Apparently the phrase originally referred to the Komodo Dragons in the Indonesian islands.

This final book in the Secret Agents trilogy is about the possible creative future that could be shaped by those who are willing to face the dragons. But it draws heavily on—and extrapolates from—recent developments in art and science. The doors have no windows. I can't know which doors will be opened or how fierce are the dragons, or treasures, behind them. I'm just grateful so many are willing to find out.

GOING WHERE THERE BE DRAGONS (EXPANDING OUR HORIZONS)

THE FUTURE. *The final pane in the triptych: a roughly drawn, unabashedly subjective map of the creative world, highlighting those places where the most daring of explorers have reported the existence of dragons, those places that at present seem most worthy of further exploration—and also are probably the most dangerous and exciting places on earth.*

GOING WHERE THERE BE DRAGONS

"The knights errant, who wandered about to clear the world of dragons and giants, never entertained the least doubt with regard to the existence of these monsters."
—David Hume (*An Enquiry Concerning Human Understanding*)

.

"The evaluation of the degree of an explorer's bravery is not contingent on real hazard, but rather on how aware he is of the infinite possibilities attendant on venturing into unknown territory."
—Hippokrites

Six billion people currently inhabit this planet. We have come a long way from when The Mother of All Mothers, Lucy Australopithecus[*], scurried along the African savannah three million years ago. It is estimated that 95 percent of all scientists are currently living, and probably almost as high an estimate could be made for artists. It is a commonplace, and true, that the increase in human knowledge is exponential. No doubt a graph of the production of art would be similarly, drastically curved upward.

.

But there are many old, scowling scholars, sitting in libraries lined with the classics, who are most definitely not impressed. Knowledge isn't wisdom, they sagely observe, nor does more art necessarily mean more good art. These old men mutter of stagnation and even more often of decadence. And often their mutterings are highly articulate and persuasive.[†]

.

One might expect that the turn of the century, and even more the beginning of a new millennium, would be a time of optimism. Certainly the beginning of a new year is greeted by most people with hope and is ushered in with good, boozy resolutions framed by cheerful expectations. But historically there is something about the

---

[*] An fascinating account of Lucy is on *The National Geographic* website:
http://news.nationalgeographic.com/news/2006/09/060920-lucy.html

[†] Jacques Barzun is but one example.

beginning of a new century that inspires in many quite the opposite vision of the future. The connotation of *fin de siecle*, as it usually is applied to the end of the 19th century, is redolent with the odour of decay. Nietzsche was perhaps the archetypal spokesman for this pessimism. Reading him, reading about "the eternal recurrence", about the decline of music as epitomized by Wagner's grandiosity, about the sick morality of pity, one gets the impression that he believed that civilization had been sliding downhill since the pre-Socratics—although, admittedly, his idea of the *Ubermensch* in *Thus Spake Zarathustra* is a model for a new, vital redefinition of humanity. And in the arts, melancholia and nostalgia were dominant colourings. "Be beautiful, be sad," intoned Baudelaire, the premier poet of *fin de siecle* decadence. And in science, this was echoed with the contemporary view that all the good stuff had been figured out and all that remained for new scientists was a rather dull tidying up.*

.

And when the scale is pushed up a significant digit, the end-of-the-world hysteria that accompanied the roll-over into 1000 AD was repeated, albeit in a more attenuated fashion, as we moved into the year 2000. For example, the "Y2K bug" panic was a comical parody of the religious panic of the last millennium change—and also extremely apropos, since technology is in some ways the new religion, and, in both cases, the high priests greatly profited by the fear-mongering.

.

But putting aside all the irrational mass reactions to what are, after all, really meaningless changes in what we scribble when we date a cheque to the utility company—there does remain the question of whether this exponential increase in art and scientific knowledge is qualitative or quantitative. Is what we're witnessing a real increase in *significant* scientific understanding? Is the sum total of all the art being produced today worth anything near as much as a single Rembrandt portrait, a Bach cantata, or a Shakespearean play?

.

It is always amazing to students of the classics to learn that so much creative genius could co-exist in such a tiny fragment of space/time as ancient Athens, with a population of about 150 thousand folk— and that includes slaves and women who were effectively disenfranchised from any direct contribution to the intellectual and

---

* Bright students thinking of a career in physics were routinely advised to look elsewhere, since all the major problems of physics had already been solved, leaving only technical housekeeping. This while Einstein was daydreaming in a patent office.

creative life of the polis. That's the size of Rockford, Illinois or Sudbury, Ontario. (Never heard of either? No surprise.) If such per capita creativity existed today, the world would be a very different, and better, place than it is.

.

The previous section reported on the present search for the secret agents of creativity. This section will continue this search into the future.

.

Ancient maps sometimes marked unknown territory—from which explorers had never returned—with the dire warning: "There be dragons." All blessings are mixed: the territories creative minds have so far opened for exploration are fraught with many hazards for those with the temerity to venture there. It is in science where this is most obvious, with the new worlds of bioengineering and nuclear science unquestionably a congenial habitat for dragons, but the artists too have ventured into strange new territory. In science, there is a clear and present danger—a physical danger. But also reason for despair, if true, is the frequently heard claim that the arts have wandered out into an arid desert and lost their way among mirages.

.

Many of the common fears of the common man (and even of those boldly going forth to explore these new territories) are mere chimerical dragons, no more real than those dragons that once were thought to inhabit the unexplored regions of the earth. But very real, never imagined, monsters may indeed lie in wait for the scientist or artist who ventures beyond the beyond. The evidence of this is overwhelming in science, where our once starring role in the grand scheme of things has been reduced to that of a bit player by our deeper understanding of the macrocosm, time, and evolution. And even in the arts we have witnessed the destruction of many cherished beliefs about our importance and the importance of what we create.

.

I, however, am more optimistic than fearful. Without discounting the dangers, and without embracing some naïve notion of progress, I believe the results of the creativity of the recent past have endowed our millennium with the potential for a spectacular renaissance in both science and art. This renaissance is only possible, however, if, as in The Renaissance, art and science resolve their petty differences and cooperate—and if the representatives of religious and political institutions don't meddle with what is outside their ken and none of their damn business. But again (perhaps naively) optimistic, I think this renaissance is actually happening.

A LOOK AT THE MAP OF TEMPTING POSSIBILITIES

Where do we venture now? This third pane in the triptych is a map—with those territories highlighted that I think most likely to be explored in the near future. The boundaries of these territories on my map are sketchy and ill-defined, but that is, of course, because they have yet to be charted. All of them, however, have one thing in common: they border on both art and science.

Science begets technology. Technological advances, in turn, clearly serve their parent science, but they also serve art sometimes less obviously.

Four 'sciences'—and the technologies associated with them—have dominated the last decades of the 20th Century. They are the physical sciences, biological science, cognitive science, and computer science. The effects generated by advances in these areas are so sweeping and have such profound implications that usually glib social commentators are left stammering, while many of the best minds of the previous generation are so befuddled by the changes that they retreat into the ranks of the Luddites.

Predicting the future is a mug's game. Futurists, such as Alvin Toffler, made predictions that demonstrated that.* Many science fiction writers have most prudently shifted the justification for their existence from predicting the future to examining the implications of present trends through futuristic exaggeration satirizing them.

So I'm not going to be so foolhardy as to predict the future; rather I will be more cautious and merely describe some of the new territories that have recently opened up for exploration and try to evaluate their potential for further exploration—and eventual colonization and exploitation. I've chosen as the most tempting areas for continued exploration those which I've been fortunate enough in my own lifetime to have witnessed the initial forays into.

One thing that all these selected territories have in common is that the early expeditions to them were accomplished by folk from both of the Two Cultures, co-operating with each other, each contributing

---

* Toffler's 1970 book *Future Shock* was a best-seller and makes interesting reading now three and a half decades later.

talents specific to their own background. Part of my optimism regarding the future of art and science is based on the undeniable collaboration of all creative individuals of every stripe in setting out to go where there be dragons.

CHOOSING A RELIABLE GUIDE

Choosing a reliable guide? There can be no guides to unexplored terrain. What is possible however, is to carry instruments, such as a compass or an astrolabe. So what comparable intellectual instrument do we have at hand? Do we possess anything to measure the 'electrical' and eclectic potential of the various paths creative activity might take in the future? I'd suggest an instrument that might cause apoplexy in many of the avant-garde in both art and science: *popularity*.

.

The new is never popular, for we are by nature conservative creatures. We like the tried and true. The new in science and the new in art have always met with resistance. In fact, much of this book is about the ways creativity transcends this innate conservatism. What is most popular is what is understood and established. How can that be an instrument to guide one into the future where so much of what seems permanently established will be overthrown?

.

It can if one makes a distinction between three types of popularity: 1) the popularity of the familiar to an audience; 2) the popularity of the moment to an audience; and 3) the popularity that transcends the moment, especially to the creators as well as the audience. Consider music as an example.

.

An example of the popularity of the familiar is the so-called "standard repertory" for so-called "classical" music concerts. You attend such a concert and, if you are a regular concert goer, most of the time, most of the music will be familiar to you. You have your Mozart and your Beethoven and your Bach. It's great stuff, and it has survived the test of time without wearing thin. But you would not expect that its—entirely justifiable—lasting popularity could offer guidance into new directions concert music might be going.

.

An example of the popularity of the moment would be the average pop rock concert. Now some of the music of these groups may eventually move into what could be called the "standard repertory" of that musical genre, as have—to give an obvious example—the compositions of The Beatles. However, most of these groups' popularity will quickly wane and their musical creations fade into obscurity, just as happened to most of the baroque, classical and romantic composers of concert music. And attending to the momentary popularity of one or another group is not going to offer any guidance into the future of music, any more than attending a Mozart concert.

.

It is the popularity that transcends the moment that may be a rough guide. By 'transcending the moment' I mean *in the relatively short term*. I also mean popularity among those who do not fall into the previous two categories—and I especially mean those who are themselves creators. In short, one might say: "What's hot with those in the know *and* has been heating up for some time!" Twentieth century concert music, during its serialist and 12-tone periods, had no popularity by any standards. But one radical break from this academic tradition in the music of Phillip Glass and Steve Reich has transcended the moment and *is* popular. It may not be popular with those who go to concerts to hear Eine kleine Nachtmusik for the fiftieth time. It may not be popular with those who want to groove in an open air concert to the Back Street Boys or to whomever is getting a lot of AM radio air time. But it *is* popular with those interested in the new, both audience and creators.

.

This is the popularity that may offer up some educated guesses as to where creative individuals will next venture. When an explorer returns from a distant land with tales and artefacts that capture the popular imagination of the adventuresome, it isn't unreasonable to assume others will follow. This isn't an infallible guide, of course, but it is the best available, and it is the one I use in this section.

AFTER MISCEGENATION: NEW HYBRIDS OR JUST MONGRELS?

Before examining the specific creative territories I believe are going to be more thoroughly explored in the near future, I want to say something about a characteristic common to most of them. It is their hybrid, or mongrel, nature.

.

Often when two species mate, the offspring are infertile. The mare and the donkey produce the infertile mule. The mule is considered smarter than either parent, than either horse or donkey, but the mule is a procreative dead-end. By now it should be obvious I don't think this analogous to the mating of science and art, except insofar as the offspring is smarter. However, there are questions—which one might even call eugenics questions—about the effects of the cross-fertilization that is the central characteristic of trends in creativity.

.

Purebred has a positive connotation. Mongrel has a negative connotation. But mixed marriages are becoming 'appallingly' common. What effect, all this miscegenation? The philosopher of aesthetics, Gotthold Lessing, maintained that great art restricted itself to what it was best suited to do, eschewing all that to which it wasn't ideally suited: Keep it clean! Reason, common belief has it, is contaminated by emotion: emotional decisions are bad decisions. The scientific enterprise has to be value-free, so aesthetic considerations have no place in it. Artists should be opposed to scientists "unweaving the rainbow." Fuzzy thinking is bad, and clear distinctions are important—but so-called 'fuzzy logic' is a new type of logic used by supposedly clear-thinking computer programmers. Science aims to see the world clearly, yet scientists, as well as artists, experiment with deranging their senses. Has the world gone mad?

.

Consider dog breeds. Consider royal lineage and attendant feeblemindedness and haemophilia. Consider the crusaders for so-called 'racial purity'. No, on second thought, let's just consider dog breeds.

.

I have two dogs. One is a purebred Labrador retriever. One is a mongrel, although the dominant genes in her lineage are Lab and Springer Spaniel. Nickel,* the purebred Lab, I worry about. The breed

---

* Nickel is so named, and unfairly, after something my mother used to say, also unfairly, of me: "You don't have to give Ken a nickel to be good; he's good for nothing."

is prone to hip dysplasia, diabetes, cancer, epilepsy, luxating patella, and retinal degeneration. And Labs are a breed with a relatively short list of highly heritable medical problems! Springer Spaniels? They too have a rather lengthy list of things to worry about, but most notable is what is called "Rage Syndrome" where a perfectly friendly Springer suddenly springs into a blind rage and attacks its owners or anyone or anything in the vicinity. The behaviour is similar to that associated with some forms of temporal lobe epilepsy in humans and may have a similar neurological cause. My mongrel (although I'm sure she'd prefer to be called hybrid) bitch (although I'm sure she'd prefer to be called *femme fatale*) was named Maggie Mae, after the Liverpool prostitute the Beatles immortalized. Do I worry about her health and the possibility that she'll snap—and then snap off some of my fingers? No. The Springer Rage Syndrome only occurs in purebreds. And all the other specific heritable medical problems associated with both Springers and Labs are largely just a problem with purebreds as well. Any vet who hasn't pledged allegiance to the flag of the American Kennel Association will tell you that mixed breeds are healthier and more robust than any purebred. My Maggie springs like a Springer and swims and retrieves like a Lab, and I have fewer worries about her than I do about my noble blue-blooded Nick.

.

Dogs, people, ideas—all are likelier healthier if they ain't pure.

SAILING OVER THE EDGE OF INHIBITION

"He who always errs on the side of caution sides with death, not life."
—Hippokrites

"I say one must be a seer, make oneself a seer. The poet makes himself a seer by an immense, long, deliberate derangement of all the senses."
—Arthur Rimbaud (*Lettre a Paul Demeny [May 15, 1871]*)

As I believe I have correctly argued, a significant creative product in either art or science is the result of both what is called divergent thinking—the generation of something new, the discovery of relationships undetected or uncreated before—*and* of convergent or critical thinking. It is claimed by many, but less than clearly established, that the 'hard part' is the divergent thinking. For the self-help types, the important question is—"how to release the creativity in us all". And this is actually a reasonable question, if one discards the silly implication that everyone is stuffed to bursting with creativity just waiting to be released. Really, what is the release mechanism for whatever little or great creativity we have bottled up inside of us?

The evidence seems to support the idea that divergent thinking is stifled by inhibition, by caution. This means that the allegedly cardinal virtue of prudence is the enemy. So how does one make the wary wild? And does being uninhibited in one's thought, one's art, one's science, necessarily mean one has to be uninhibited in every other aspect of one's life? All the brilliant people whose lack of impulse control has ruined their lives suggest this may indeed sometimes be true, but that does not prove it so. Some certainly can cordon off their wild side to an arena for artistic and/or scientific explorations, and keep their personal lives safely behind the cordon in a place of moderation and social conformity.

It may be that moderation in all things (as was the Athenian ideal for the good life) is a path to a long and contented life. However, excess in all things has been the path taken by many of the most genuinely creative. That this uncontrolled excess has in many cases destroyed their lives seems to be a risk many were willing to take—or couldn't avoid. The recent advances in science and technology and social tolerance have made possible excesses undreamt of before. (At least sometimes and some places, for intolerance has become extreme as well.)

.

Extremism now seems more admired than moderation, and what was once considered extreme is now considered moderate. The popularity of 'extreme sports' is indicative of this trend. The already extreme combative sport of professional boxing (based on the polite Marquess of Queensberry Rules) is now considered tame and boring, and the new rage is Extreme Fighting which has been called "human cock-fighting" by its detractors. The Ultimate Fighting Championship is a Mixed Martial Arts (MMA) that attracts more fans than conventional boxing. The two opponents are put in a chain link cage to fight it out, using virtually whatever brutal methods they want to beat their opponent into submission.* Two men go in. One walks out. The other is carried out—or hobbles out, bloody and beaten. Gladiatorial? Yes. So not new? Yes. But this isn't ancient Rome. This is a time where governmental nanny-state obsession with safety has reached epidemic proportions.† It is remarkable—if not entirely surprising—that total disregard for personal safety by individuals is seen as admirable, and activities involving unnecessary risk have so much popularity.

.

Extremes are interesting. Extremists are interesting. Extreme opinions are interesting. And all become especially interesting when social and political institutions are becoming more and more restrictive. And all forms of extremism are especially interesting to creative individuals. Some will delve into this world of extremes just as observers, but some will be lured in as participants. This has always been the case. In many ways this world of extremes has become more

---

* There are a *few* rules such as not being permitted to put one's finger in any orifice or open wound. (So you can't gouge your opponent's eye out or rip open a wound—or goose him!)

† Examples abound. For example, in Ontario, there is a movement to remove playground equipment because occasionally some child has fallen from a set of monkey bars or swing set and hurt himself. Seat belt legislation and bicycle helmet legislation are other examples of this. Meanwhile, people flock to see formula-one races and BMX bike races, where the chance of serious injury or even death is substantial.

extreme than ever before, so one can expect more from our creative people—and more casualties.

THOSE OLD FAVOURITES: SEX AND DEATH

In 1973 the Canadian poet Al Purdy published a collection of verses called *Sex And Death* and in doing so scooped all future writers with *the* ultimate book title: the one that sums up the central concerns of writers and artists and the rest of humanity. It baldly states what is implied in myriad other book titles: this is about sex and death. That's a hook to grab everyone, for everyone is obsessed with both more than anything else in life or literature. Most tragedies deal with these two themes, often intertwined. Most comedies do so as well, as do most jokes: "The difference between sex and death is that with death you can do it alone and no one is going to make fun of you."*

.

Objections?! What about love? What about pain and suffering?† These are but sub-categories of sex and death. Okay, I know I'm exaggerating slightly for effect, but, really, most literature that deals with love is not about maternal love or brotherly love or deep friendship: it is about the sexually-based attraction between two individuals. And pain and suffering usually are but the drama that culminates in death real or avoided.

.

Sex—and love. Again acknowledging that any extrapolation from current trends to the future is a risky business, it can't be denied that the dramatic social changes in sexual relationships (and reflected in art) that have occurred in the last fifty years do not seem to be levelling off. When I was entering adolescent hormonal surge the idea of same-sex marriage was just a bizarre theme of some science fiction works. Just a few years before a virtually fool-proof contraceptive in the form of a pill was almost inconceivable (pun intended). Even a few decades ago, who really believed some researcher could rise (ahem) to the challenge of finding a cure for impotence in the form of another magic pill? And what about the acceptance of the idea that even this magic pill was unnecessary, because one could bypass the pill-taker: i.e., the idea of women being able to purchase mechanical devices that do a better job of delivering them to orgasm than their lovers? And then there is *in vitro* fertilization and the possibility of parthenogenesis and cloning. The ways these things have affected

---

* Woody Allen (quoted in *New York Tribune*, 1975)

† What about taxes! How come no one writes about that?

social interaction are profound, and the 'sexy' subject of study by historians of contemporary change.

Science is the not-so-secret agent behind the dramatic changes in sexual attitudes and behaviour of the last half-century, while art has both reflected these changes and influenced them. Two core changes in attitude are those regarding homosexuals and women.

Homosexuality is hardly a new phenomenon—nor is it, as far as one can tell, more prevalent now than in the past, despite what the religious fundamentalists claim. Just as there seems to be a disproportionate number of great artists with bipolar disorder, the numbers of homosexuals in the artistic community seems considerably greater than in the general population.* In both cases this is difficult to confirm but especially so for homosexuality, which still seems too controversial for unbiased and accurate study. But assuming that is a valid generalization, what is of interest here is not the sexual orientation of the artist, but the way it is reflected in his (or her) art. What was implicit has been made explicit. Sex is so central to our self-definition that it *must* influence what we create, but how explicitly it does this is what has changed.

Tchaikovsky's "Romeo and Juliet Overture" is about as explicitly erotic as a work of absolute music can possibly be.† But the erotic impulse that inspired it was not heterosexual, for this great romantic composer most definitely was not. (This is a fact the Soviets attempted to suppress, for like the Nazis, they believed homosexuals to be degenerates and Tchaikovsky was a national hero.) And even the original author of the play that is archetypal about young heterosexual love just may have been gay.‡

In literature Oscar Wilde is exemplary of the relationship between the artist's sexuality and its artistic expression in the not so distant past. Like Tchaikovsky, his sexual preferences landed him in a lot of

---

* I suppose I have to say, emphasize, that this comparison is *not* meant to imply that homosexuality is a mental illness.

† It comes highly recommended as 'mood music' for seductions.

‡ So little is know about Shakespeare the man, and the way his own personality is invisible in his work, allows considerable speculation. His sonnets, because lyric poems and so assumed to have autobiographical elements, have been cited as circumstantial evidence of his homosexuality—or at least bisexuality.

trouble, although he was far more open about his predilections than was the composer. Wilde's writings are never explicitly homosexual in theme or content, but at the risk of being accused of stereotyping homosexuals, it seems to me his famous wit is distinctly—*bitchy*.

.

Contrast his wit with the also oft quoted H.L. Mencken. Both are brilliant at wounding their opponents, but the difference in tone and approach is obvious. Mencken bludgeons; Wilde jabs.

.

- Wilde: "A little sincerity is a dangerous thing, and a great deal of it is absolutely fatal."
- Mencken: "All men are frauds. The only difference between them is that some admit it. I myself deny it."
- Wilde: "America is the only country that went from barbarism to decadence without civilization in between."
- Mencken: "In the United States, doing good has come to be, like patriotism, a favourite device of persons with something to sell."
- Wilde: "A man can be happy with any woman as long as he does not love her."
- Mencken: "For it is mutual trust, even more than mutual interest that holds human associations together. Our friends seldom profit us but they make us feel safe... Marriage is a scheme to accomplish exactly that same end."
- Wilde: "Always forgive your enemies; nothing annoys them so much."
- Mencken: "Every normal man must be tempted at times to spit on his hands, hoist the black flag, and begin to slit throats."

.

Allegedly Wilde's last words in a shabby Paris hotel room were "Either this wallpaper goes—or I do." Reading these words, how can one not see the exaggerated effeminate gesticulation? Mencken suffered a stroke and so was mentally incapacitated on his death bed, but surely had he had his wits about him he would have said something like what W.C. Fields had inscribed on his tombstone: "All things considered, I'd rather be in Philadelphia."

.

Somewhere mid-Twentieth Century, the implicit effects of artists' sexuality on style became explicit in the content of their creations. The infelicitous phrase "coming out of the closet" is what artists did in not only their lives but in their art. I remember, back in the early sixties, the controversy at a Chicago community college when a professor assigned James Baldwin's *Another Country*—which explicitly deals with sexual relationships of every imaginable permutation and combination—as reading for his course. Baldwin had previously published *Giovanni's Room* (1956) which is a novel about homosexual love, but it was *Another Country* that seemed to attract the most attention. Obviously, any literary scholar can cite precedents for explicit dealing with homosexual themes dating back at least as far as Sappho; I'm only suggesting that it was sometime in the sixties that homosexuality started to become explicit in the arts and widely accepted by the artistic community—if not yet the mainstream audience.

.

One has to wonder what Oscar Wilde's response would have been to *Another Country*. What Michelangelo would have thought of Mapplethorpe's sadomasochist and homo-erotic photographic art? Surely they would be shocked at the explicitness—although presumably impressed that such expression was possible. (I have no idea if they would be impressed with the works as works of art.)

.

I recently watched the film *Brokeback Mountain*, which won three Oscars and numerous other mainstream awards. It is a bittersweet, not too sentimental, love story about two bisexual cowboys which seems to have moved heterosexual audiences as much as any good, well-crafted love story would have. So it seems reasonable to expect that explicit exploration of formerly taboo sexual subjects will extend beyond the now almost passé one of homosexuality. Such exploration by artists has been going on for a long time, but only recently are the explorers' reports of their expeditions being made accessible to the general public—and treated with interest, rather than disgust.

.

The other core change in attitude is even more radical, for unlike homosexuality, it seems to almost never have had any kind of wide social acceptance. Of course I'm referring to the attitudes toward half of our species: women. Many feminists have made an academic industry of tallying up past abuses and injustices and stupidities, so nothing I could possibly say would be anything more than a commonplace observation.* What most obviously is germane to future developments in the arts and sciences is that it far easier now

---

* Of course I'm not implying there aren't still abuses and injustices and prejudice.

for creative women to Actualize their potential. And it is also easier for them to explore their sexuality.

.

Flaubert's 1857 masterpiece *Madame Bovary* was considered scandalous and the novelist charged with obscenity*, because it was a sympathetic portrayal of an adulteress. Compare it to Erica Jong's 1973 best-selling novel *Fear of Flying*, which is not only far more explicit but, more importantly, authored by a woman obviously unashamed of her own sexual adventures.† As with Baldwin what is crucial is not the actual explicitness of the language‡, but the explicit acceptance of the social *legitimacy* of the sex—in Baldwin's case homosexuality and in Jong's case female sexuality.

.

That creative men and women often were sexual adventurers is well established. What *is* new and foreshadows even greater change is the acceptance of all the varieties of sexual experience as legitimate and almost mainstream—at least in the educated audience for creative endeavours. The effect in art is an expansion of the canvas on which the artist may work without fear of censure or censorship. The effect in science is that the sex or the sexual behaviours of the scientists is less and less an obstacle to their doing their creative work.

.

Death—and suffering unto. Here too science is the not-so-secret agent behind the dramatic changes in attitudes toward our other central obsession. Science has reduced suffering dramatically and science has the potential to increase suffering even more dramatically; for example, respectively, anaesthesia and biological or nuclear weapons. Science has also had a profound effect on attitudes about death and dying, not to mention increasing the life expectancy, and so creative output, of our most talented.

.

A good case could be made for advances in medical science having had the greatest positive effect of all scientific discoveries on the lives of all human beings. It is appalling to realize that the philosopher

---

* He was acquitted, and the book sold well. Nothing, like being 'banned in Boston' (or almost banned in Paris) to sell books.

† Erica Jong denies that the novel is autobiographical but admits to autobiographical elements.

‡ For example, Jong coined the phrase "zipless fuck" in the novel, and of course Flaubert's writings contain nothing quite so blunt.

Thomas Hobbes' observation that the majority of human lives have always been "nasty, brutish, and short" is indubitably true. And this just doesn't apply to ancient or medieval times or countries like Sierre Leone or American inner city ghettos. One study I found showed that in Massachusetts only a century and a half ago (1850) the life expectancy at birth of a white male was 38 years. In 2004 it was 70 years. This dramatic increase is partially a result of decreased infant mortality rates. Even a ten year old in those days, who had survived birth and childhood diseases, could only expect to live to the less than ripe old age of 48. Furthermore, he or she was likely to have a pretty low quality of life because of ailments that were not easily cured or even ameliorated, as they are today. Effective analgesics and anaesthetics are a relatively recent development. In fact surgery without any kind of anaesthetic was common throughout much of the 19th century. It is a trivial example when compared to major surgery such as amputations, but even I can remember going to the dentist as a child and having filings and tooth extractions done without anaesthetic.

.

It is depressing to contemplate the ailments that inflicted so very many of the great creative minds over the centuries and to think about their short lives that seem so tragic for them and for us who can fantasize about what would have been produced had they lived a reasonable span. Mozart didn't reach forty, and he suffered from numerous ailments all his brief life that would have made most of us incapable of work of any kind—never mind creative work. Bach, who did live to the incredible (for the time) age of sixty-five, buried nine of his thirteen children before his own passing. The short lives of the great Romantic poets Byron, Keats, and Shelley* may be tragic, but they weren't atypical of their less gifted contemporaries.

.

Other advances in biological science and medical application of this knowledge have raised ethical issues inconceivable a hundred years ago. Immortality has always been a theme in art, but scientific advances have given it real world plausibility. Of course I don't think we are ever going to 'cure' the aging process, but greatly extended life expectancy is no more than an extrapolation from what has been accomplished already. The implications of this are speculated on by science fiction writers, but they are also already having real effects on

---

* Byron at thirty-six. Keats at twenty-six. Shelley at thirty.

social policy, as in concerns about paying for health care and social security for an aging (refusing to die) population.

.

Consider the moral issues involved with euthanasia, organ transplants, sex change operations, stem cell research, cloning, genetic engineering, abortion based on in-uterine identification of disease or even creative potential, cryogenics, human memory being interfaced with digital storage systems, drug or psychosurgery treatments that affect personality and intellectual ability. Artists have many, many new themes for plausible exploration and extrapolation, and scientists are daily increasing these possibilities. There is not room in this book for me to even begin to speculate about these changes. The scientists will make them possible, and the artists and philosophers will be busy dealing with them. That I can say with assurance.

.

I can also say with some confidence that suffering and death will remain a central concern of both artists and scientists. It is an eternal theme, but the new variations on it we can expect are mind boggling.

.

Before moving on to the next new territory, there is an issue related to our all-too-human obsession with sex and death deserving of a few comments, especially since it seems to be of such great concern to the self-appointed guardians of public morality and aesthetic purism. It is the apparent increase in extremism in raw sensory stimulation. Music gets played at louder and louder decibel levels. Depictions of sex and death become more and more graphic and explicit in film and books. Artists seem to strive more for shocking their audiences than with producing works of lasting aesthetic interest. Perhaps most exemplary of this trend is 'Reality TV' shows where there is no real artistry but only partially scripted raw reality which, even at its more banal or vicious, is shocking because it exposes what is usually not public.

.

To use shock tactics myself, let me say something about 'snuff films', those films that allegedly record the real sexual abuse and subsequent murder of a woman. As far as I can ascertain all the films that purportedly record this atrocity are faked, but I don't doubt that someone somewhere has made—or will make—such a film. Nor does the misanthropic cynic in me doubt that there is a wide audience for such a film, for demonstrably there is a distressingly large audience for 'reality' videos of child-abuse pornography. Is this a new territory being 'explored' by creative individuals? That is a rhetorical question. This of course has nothing to do with radical creativity or creativity of any sort.

.

The news is constantly reporting on 'performance artists' who kill animals or mutilate themselves to make some point—although of course their real motive is just to draw attention to themselves. Hollywood jacks up the amount and realism of violence and mayhem every year—and audiences become more and more jaded. Many of us have our eyes glaze over during the scenes that are intended to rock and shock us.

.

The horrific violence in Sophocles' *Oedipus* all occurs offstage and is much more aesthetically effective than if some actor with the help of special effects could seem to put his own eyes out on stage. The violence in Stanley Kubrick's *Clockwork Orange* is more disturbing than that in any of the films by such cinema gurus of violence as Sam Peckinpah (or Quentin Tarantino or David Cronenberg) precisely because it isn't realistic: it is choreographed.

.

So one current trend in aesthetic expression of the eternal themes of sex and death that I think will level off is the movement toward extremist realism. One function of art may be to shock us out of complacency and see the world anew, but that shock has to be to our way of thinking, not merely the momentary primitive shock delivered by taboo words or gory images or outrageous behaviour called 'performance art'.

.

Strangely, it is scientists, more than artists, who lately seem to be shocking us most effectively, although that is not their intention. The scientific discoveries about the nature of our sexuality, about the neurological correlates of our suffering (as well as our happiness), about aging and death (as well as about the nature of life itself)— these are far more profound in their effect on our view of things than the latest cause célèbre about explicit sex in a film or an art gallery installation that offends some people's religious beliefs.

THE CHEMICAL SOLUTION

The use of chemicals that affect our brains is as old as recorded history. The usually desired effect is cutting of the bonds of inhibition, so stimulating us to action. Inebriants and stimulants, these are swords to cut the bonds that bind creativity.

.

The first, and still most popular, drug to effect this liberation is alcohol. One reputable scholarly theory as to why our ancestors changed from hunter-gatherers to home-bodies—and developed an agrarian culture—is because they had to wait around for the beer to ferment. Many claim that the making of beer even preceded the making of bread, many dating its origin to the 7th millennium B.C. Extant documents indisputably confirm that the ancient inhabitants of Egypt and Mesopotamia were beer drinkers. So zymology or zymurgy was one of the first sciences—if last in the dictionary.

.

I should say that the word 'beer' can be misleading. I doubt anyone first tasting Sumerian beer would easily identify it as a beer.* Beer is a generic term for an alcoholic beverage created by fermentation of grain. (Sake, for example, is actually rice beer—not rice wine, as commonly assumed.) Wine is an alcoholic beverage created by the fermentation of grapes or other fruits. Because the yeast needed for fermentation is killed by it own offspring (alcohol) if it becomes too concentrated, so-called 'hard liquor' didn't come onto the scene until distillation was rediscovered in the 8th Century by the Arabs. Only in the 12th Century did the technique become popular in Europe for the production of stronger alcoholic drinks.†

.

The simple, somewhat unsavoury, fact is that almost anything that rots can ferment and produce alcohol, and so it isn't really surprising that our ancestors discovered these alcoholic by-products and their mood altering effects. From Sumerians with their sweet-bread beer through the palm-wine drinkers of Africa to the absinthe drinkers at *fin de siècle* Paris cafes on up to Budweiser chuggers at a Phi Beta Kappa party, the presence of alcohol has been—and still is—virtually

---

* Based on discovered recipes, versions of it have been brewed, one even commercially. Apparently it is very sweet and except for its alcohol content bears little taste resemblance to what we call beer. It has no hops and certainly isn't a wonderful pilsner.

† Certainly a time when a 'stiff' drink was often needed.

ubiquitous in human culture. It is true that repressed and repressive cultures and subcultures try to enforce abstinence, but they are the exception and sometimes are quite tolerant of the use of other drugs such as hashish.

.

"Everybody must get stoned!" sang Bob Dylan. Stoned, smashed, hammered, whatever. This was The Sixties Message: the world would be better if everyone used chemicals to cut the bonds that bound them. Alcohol wasn't the most hip brain-altering chemical then, but marijuana ("stoners") or drinkers ("juicers") were both seen to be working together to effect a global change in consciousness—to free up the creative juices that would right the world's innumerable wrongs. It is an appealing message, for what could be more appealing than doing what makes you feel good while feeling you are somehow *doing good* by feeling good?

.

Before even considering the bewildering variety of mind-altering substances now available—and which may or may not open the doors of perception and creativity—it is worthwhile to look briefly at the history of this all-too-human tradition—and addiction.

.

Alcoholic beverages (and to a lesser extent other drugs) have always been associated with ceremony and celebration, be it religious or secular. Examples are endless: Every event from ancient Greek Dionysian rites, Roman orgies, weddings and wakes, parties and partings, even the Christian sacrament where one drinks wine as symbolic of drinking the blood of Christ* all have traditionally involved loosening our inhibitions through the infusion of ethyl alcohol into our brains. With our inhibitions suppressed, we are more open to experience, more susceptible to suggestion, more expressive of what propriety and sobriety normally keeps locked inside us.

.

In a place where authorities fear this unleashed expressiveness, every attempt is made to suppress the use of alcohol and all other disinhibitory drugs. Religious fundamentalists everywhere demonize

---

* This is a particularly disconcerting example, for it has such a pagan, even savage quality to it that few Christians seem to note. Conservative Protestants seem a little more discomfited by it. The Presbyterian church I was sent to as a child used Welch's Grape juice instead of wine, which I don't believe has any psychoactive ingredients, while the Lutheran church my wife was sent to as a child used white wine—which has lower alcohol content but seems somehow to miss the point symbolically.

alcohol for this reason, Islamist and Christian fundamentalists being the most obvious examples. Of course, in the end, suppression doesn't work. The United States offers good evidence of the futility—and the ultimately adverse social effects of such attempts. Prohibition lasted thirteen years, and alcoholism increased dramatically during this time. The United States found itself at loose ends when the Cold War ended in a whimper. The tendency of demagogues to search out a panacea for social and economic ills led to the creation of a new Cold War—the "War On Drugs".* In response, drug abuse and drug-related crimes have continued to soar in a positively correlated relationship with the efforts in this war, just as organized crime grew out of Prohibition.

Although this is not the place to editorialize about the absurdity of "Zero Tolerance" and all the other nastiness associated with this new 'war', one important effect relevant here is that politics confounds any objective attempt to understand the role chemical intervention has had, or might have in the future, on creativity—just as politics and ideology confounds any objective evaluation of the relative role of nature and nurture. It is asking for trouble to even suggest that the inspiration of jazz musicians might be increased by inhaling smoke from a legally banned plant. Of course it is also disturbing to many artists themselves to suggest that the products of their labours and rigorous training could just as easily be produced by the introduction into someone's body of a psychoactive chemical.

Before considering the future role chemistry might play in creativity, it makes sense to look at the evidence from the past regarding the use of such commonplace disinhibitors and stimulants as alcohol and beer. And once again doing some preliminary taxonomy is worthwhile. For purposes of discussion here psychoactive drugs can be roughly sorted into three categories: 1) disinhibitory; 2) stimulants; and 3) therapeutic.

These categories are not congruent with the conventional, more neurologically-based, four-part sorting into stimulant, depressant, hallucinogenic, and therapeutic; nor are they mutually exclusive categories. But they are the most useful taxonomy for considering the role of drugs in creative endeavour.

---

* While still going strong, the U.S. government now is splitting its limited resources by simultaneously waging a "War On Terror". And here too, the "war on" has only increased the numbers of its enemy.

Disinhibitory. *One must be uninhibited to create.* No doubt Ninkasi (the Sumerian goddess of beer) was precursor to the Greek Muses. Nietzsche, as usual, put it best: "For art to exist, for any sort of aesthetic activity or perception to exist, a certain physiological precondition is indispensable: *intoxication.*"

Freud was right about one thing: we need our id (our libido, our knuckles-to-the-ground urge for instant gratification) to be kept under control, or else we will do great harm to ourselves and others. Inhibition makes civilization possible. But since it is also true that one has to be uninhibited to create something new, there is an inevitable conflict between the creative individual who is responsible for development of civilization and the properly controlled and inhibited individuals that maintain and sustain its existence—between the staid and responsible trustees of a gallery and the crazy, rude artists that exhibit there. Yet both are essential.

More importantly, even within the creator, both are needed: the uninhibited creation must be refined and defined by the more cool-headed, critical part of the artist or scientist. Yes, wild ideas must be welcomed, but they are useless until they can be evaluated, and bad ones discarded while the ones with potential refined. And, ah, there's the rub!

The artist or scientist must be Dionysian first and then Apollonian afterwards. This is asking a lot, asking for a lot of inconsistency (or flexibility) in personality: go forth and be mad and passionate, but then throw some mental switch and become rational and objective. Enter the chemical assistant to the rescue.

Generally, the hard part, strangely enough, is to be uninhibited—not to be self-indulgent. All properly raised children are taught to conform, be agreeable: 'behave themselves'.* That this is so thoroughly inculcated in us is the reason we can live together in civilized communities. So, contrary to the popular conception of the artist as naturally uninhibited, most artists are like everyone else in finding the Apollonian, the critical, the rational, the conformist and

---

*Richard Dawkins in *The God Delusion* goes so far as to suggest that possibly the fundamental Darwinian principle of survival of the fittest has resulted in an innate willingness to listen to our elders injunctions. Some may be pointless but harmless. Others are good advice for survival in a hostile world.

agreeable side of their personality dominant. It is this side that needs at least temporary suppression to release the Dionysian, uninhibited side to create something new.

.

Time to have a beer. Lift a glass to Ninkasi!

.

Consider the writer who sits down at his typewriter or word-processor. Assuming he isn't in some frenzy from lust or rage or some other passion, he probably risks the dreaded writers' block. He needs to 'let go'. And just as a few drinks at any social gathering allow us to drop our inhibitions, a bit of booze just might help getting the literary juices flowing. The work produced can always be evaluated, edited, censored, in the cold and sober light of morning—unlike those 'inappropriate' remarks we made at the office party or sent off in an angry late night email.

.

Let's keep our focus on the writer as representative of this issue. Writers are notorious for being serious drinkers. Is this justified? I think it is. Like many other stereotypes that the politically correct find distasteful, this one has more than a grain (pun intended) of truth: writers seem to like their booze. The author of the fabulous ribald classic, Rabelais, was frank about it: "I drink no more than a sponge." Not all writers, of course, endorse drinking, but then there is nothing *all* writers endorse. The author of *In Praise of Older Women*, Stephen Vizinczey, included "Don't drink alcohol!" as one of his list of ten rules for writers. (But then he was a one-book wonder, and it was women that intoxicated him.)

.

So while I know of no carefully controlled scientific study on this, I feel confident in saying—on the basis of personal experience, anecdotal evidence, extensive reading of authors' biographies and even psychological research such as Ludwig's*—that writers have traditionally used alcohol to loosen their tongue. The ancient Greek dramatist Aristophanes proclaimed (in *Knights*) "Quickly, bring me a beaker of wine, so that I may wet my mind and say something clever." Horace proclaimed "Now is the time for drinking, now the time to beat the earth with unfettered foot." The great romantic poet and rogue Lord Byron proclaimed "Man, being reasonable, must get drunk; the best of life is but intoxication!" Thomas Wolfe, author of the American classics *Look Homeward Angel* and *You Can't go Home*

---

* Previously cited.

Again– said alcohol was a "life-enhancing genie." Mark Twain remarked that "sometimes too much to drink is barely enough." John Berryman agreed: "Something can be said for sobriety but very little."

.

Of course alcohol can be a mixed drink, a mixed blessing. "Be wary of strong drink. It can make you shoot at tax collectors... and miss." Robert Heinlein warned. But a more serious remark by C.K. Chesterton really does express the double-edged sword nature of drink (or almost any psychochemical intervention): "No animal ever invented anything as bad as drunkenness—or so good as drink."

.

The sad saga of the brilliant poet Dylan Thomas drinking himself to death is almost archetypal of the writer's life. Even a modest list of good and great writers who fell into the bottle would take many pages. The problem with drink, with chemically induced disinhibition, with loss of inhibition in general, is that it just feels too good to be restricted to the role of a means to the end of creativity. We may have to let the beast out of the cage to apprehend the world beyond, but he isn't easily lured back in. Freedom is addictive, even, or especially, irresponsible freedom.

.

What is frightening is that alcohol is probably the most *innocuous* of the chemical disinhibitors, although fans of marijuana might reasonably dispute this. But certainly the 'recreational' drug revolution opens up new and far more dangerous territory than did beers and reefers. "Anyone who remembers The Sixties wasn't there." So sayeth the survivors of the psychedelic revolution, but some of those that did survive just have little coherent to say.

.

The 'discovery' of hallucinogenic drugs was initially hailed as a discovery of magic pills which would open "The Doors Of Perception". This, of course, is the title of Aldous Huxley's 1954 book about his experiments with mescaline that he felt opened doors into deeper reality, somewhat like the experience of the man who left Plato's famous cave of shadows. He wrote that "The man who comes back through the Door in the Wall will never be quite the same as the man who went out. He will be wiser but less sure, happier but less self-satisfied, humbler in acknowledging his ignorance yet better equipped to understand the relationship of words to things, of systematic reasoning to the unfathomable mystery which it tries, forever vainly, to comprehend." It is ironic, even paradoxical, that Huxley is also the author of the classic dystopian novel *Brave New World* (written twenty years earlier) that features a drug called "soma" as one of the tools of control in a totalitarian state that, by offering chemical pleasures, keeps its doped citizens docile. It is interesting to note that after his experience with mescaline, Huxley never produced another significant literary work* and became involved with the most flakey of movements, including the Esalen "human potential" movement and psychic 'research'. He took LSD on his death bed, presumably to travel beyond the beyond both chemically and physically.

The apparently brain-addled post-mescaline Huxley seems an unlikely role model, but the dire warnings of *Brave New World* were not what the youth of The Sixties attended to—rather it was the promises of instant spiritual enlightenment offered up in his *Doors Of Perception*. The famous rock group, *The Doors*, took their name from Huxley's book, and their charismatic lead singer, Jim Morrison, apparently also took his attitude toward drugs from this book, the drugs that eventually killed him at age 27.†

Morrison is still idolized as a poet and visionary, and travelling to his gravesite at the Père Lachaise Cemetery in Paris is a pilgrimage many aging hippies still make. That Morrison was a creative and a talented musician is undeniable, although few serious readers of poetry would credit him with much talent in that domain. How much of his creativity can be attributed to his experimentation with disinhibitory drugs? A good guess would be that it was about the same as Dylan Thomas's creativity was aided and abetted by his preferred disinhibitory drug: alcohol.‡

The critical historical difference here is two-fold: 1) that new, more potent disinhibitory drugs were becoming available; and 2) that these new psychoactive chemicals were imbuing us with powers greater than mere disinhibition. Few had believed that just getting drunk or stoned would automatically make you creative—never mind making you more 'spiritual' and able to grasp the meaning of life. But The

---

*He published another novel, *Island*, which was not a biting dystopian satire, but rather a lame and silly utopian novel where psychoactive drugs are the quick route to enlightenment and the good life.

†I know, I know, actually the CIA murdered him or he is still living in India! Conspiracy theories abound.

‡ Morrison, it should be said, was a notorious drinker as well.

Sixties were packed to overflowing with young people who believed that insight and creativity came on an acid tab.

.

The so-called psychedelic art produced during this period has not stood the test of time. Most of the visual art is an unmitigated disaster—which is odd, given that the new drugs real primary effect was on visual perception. The literature produced is less numerous (short bad poetry excepted), probably because it requires more effort, has a more sober analytic aspect. LSD loving Ken Kesey*, of the Merry Pranksters fame, might be cited as a notable exception, but the fact is that his two masterful novels (*One Flew Over The Cuckoo's Nest* and *Sometimes A Great Notion*) were written before he devoted his life to taking hallucinogens as a lifestyle and took to the road in his famous school bus named "Furthur". It is true that he had experimented with psychoactive drugs long before he wrote his masterpieces†, but again this only points to both the potential usefulness of disinhibitory drugs *and* the potential danger of them changing from means to an end in themselves.

.

Things have 'progressed' greatly since the Merry Prankster, Haight-Ashbury, and hippie times. DuPont's advertising slogan from 1939 seemed prescient: "Better Living Through Chemistry!"‡ There are so many new drugs available—a virtual pharmaceutical cornucopia! Is it time for Thanksgiving? Or is it not so much a cornucopia as a Pandora's Box? I would venture to say both.

.

On the asset side of the ledger, there is the refinement of these psychoactive chemicals. As the variety of disinhibitory drugs increases, so too does their specificity of effects. Alcohol works to lower inhibition, but neurologically it is classified as a depressant. You may think more creatively after a few drinks, but you think less clearly, for cortical processing is being depressed. Alcohol draws on the brain's canvas with broad strokes.

.

---

* Tom Wolfe's book, *Electric Kool-Aid Acid Test*, is a wonderful book *about* Kesey's antics and his experimentation with psychedelics. But Wolfe was writing as an observer, not a participant.

† In 1959 Kesey had volunteered to take part in a CIA-financed study on the effects of such psychoactive drugs as LSD and mescaline.

‡ The phrase "through chemistry" was dropped in the eighties and has recently been replaced with "through the miracles of science". Wonder why.

We already live in an age of designer drugs, drugs whose effects are far more specific than alcohol or, for that matter, any of the 'traditional' disinhibitory drugs such as marijuana or the opiates or the LSD and the natural hallucinogens such as mescaline and peyote. The term "designer drugs" was originally applied to variants of existing psychoactive substances that were developed to avoid illegality by not quite matching the defined chemical characteristics of the banned substances. Eventually, however, it came to be applied to drugs (or precise combinations of drugs administered together) which have been designed to have highly specific effects. One entirely legal example that has recently hit the market is a beer produced by Anheuser-Busch that contains caffeine. It combines the stimulant effect (to be discussed shortly) with the disinhibitory effect of alcohol.

.

As for more sophisticated and potent designer drugs, the pharmacologist Alexander "Sasha" Shulgin was the guru. Shulgin was a reputable scientist with a Ph.D. in Biochemistry from the University of California who worked for (as a research director) BioRad Labs and then for Dow Chemicals. Eventually he became an independent researcher (with a lab in his backyard shed) and collaborated with the infamous U.S. Drug Enforcement Agency (which licensed and funded his research and gave him awards for his contribution to their efforts). But then in 1994 the DEA turned on him and raided his lab. This was shortly after the publication of two books he co-authored with his wife: *PiHKAL* and *TiHKAL*. These books allegedly were being used as 'cookbooks' in many of the drug-dealer labs raided by the DEA.

.

Pandora's box can't be repacked. It is reasonable to predict that more and more 'refined' and 'designed' drugs will be produced. And it is also reasonable to expect that people who wish to enhance their creativity will be among the most enthusiastic experimenters with these drugs. The biographies of the creative offer up plenty of evidence of writers, musicians, artists, and scientists manifesting their 'openness to experience' in this way.

.

The unanswered question is whether this development will, overall, have positive or negative effects. The answer to this question depends on whether or not it really is possible to develop psychoactive substances that can offer up the disinhibitory effects that release creativity without any deleterious side effects such as aggression, addiction, or the snapping of the bonds of social and personal responsibility. "Never say never!" is good advice to any would-be predictor of the future, but I have my doubts about such optimistic predictions. Every silver cloud has a dark lining.

.

It is difficult to imagine that disinhibition can be chemically focused to do a precision strike on a single target. Let's blast out that writers' block, but we'll have no collateral damage! Let's free up those dark fantasies but not let loose into the streets any other dark, asocial entities!

.

Potency may be one critical factor here. At least that is what history suggests. Coca leaves as a stimulant were—and still are—integrated into certain South American cultures with minimal deleterious effects, but cocaine is much more dangerous—and its further refinement into crack cocaine has done incredible harm. The ingestion of alcoholic beverages made by fermentation certainly has been the source of much misery, but the deleterious effects of alcohol were exponentially aggravated by the application of distillation in its production, which greatly increased potency.

.

They are inextricably linked, creativity and reduced inhibitions. Some artists and scientists manage to free up the demons that are useful aids to their endeavours, while still keeping those less useful demons on a short leash. Probably that won't change. The danger is that more will be considered better, and the more on the dark side will wipe out the more on the light side.

.

Stimulants. *One must be stimulated to create.* Those most stimulated produced the most. The incredibly prolific novelist Balzac allegedly died from caffeine poisoning.* And the equally prolific composer Bach had as his most prized possession his coffee maker. And then there is the great mathematician, Erdös, mentioned earlier, who moved on from a voracious caffeine addiction to cocaine.

.

Consider Honoré de Balzac, author of the great 95 volume *La Comédie Humaine*, who died at the age of 49. That he was hopelessly addicted to caffeine is well-known: he, at the end of his life, had taken to consuming dry coffee grounds to get his fix. Balzac's commitment to his art is almost unrivalled. Allegedly he wrote or revised proofs for up to fifteen hours a day, and did this without

---

* Technically his death was from a perforated stomach ulcer, which undoubtedly was the result of his excessive consumption of strong, black coffee.

giving up his social life in Parisian society, from which he gathered much of his material. When, one asks, did he sleep? Whenever, but certainly not very much. And he had his beloved coffee muse to make sure of that.

.

Johann Sebastian Bach singled out in his will his beloved coffee maker as his most prized possession. Coffee in his time had a status similar to marijuana now: it was actually illegal, a banned substance, albeit one used by many. Coffee was cool. Coffee was hip. Coffee shops had a status similar to that of the speakeasy during prohibition.* And like now, there were many finks around who'd run to the authorities if they smelled something naughty—which then was the wonderful rich aroma of coffee and now is the pungent smoke from a cigarette or a joint. Bach even wrote an amusing "Coffee Cantata" wherein the heroine will only agree to marriage if acceptance of her right to indulge her addiction to coffee is written into the contract. Bach produced well over a thousand works. The standard catalogue (BMV) goes up to 1,126, but some would estimate that almost half again as many works have been lost to posterity.

.

Coffee is one thing, cigarettes another. Nicotine, like coffee, is a neurological stimulant, and it increases levels of adrenalin and the neurotransmitter dopamine, both of which are associated with arousal, increased energy levels, and pleasure. It seems to directly affect the pleasure centre circuits of the brain. Natives in the Americas used tobacco, apparently with restraint, long before Europeans arrived, but it quickly caught on with the new-comers and was exported back to Europe where its popularity grew rapidly. Its use by the creators in society has always been and is still greater than in the general population—this despite the fact that creative folk are supposedly brighter than average and the frightening health risks associated with smoking are indisputable. The smoky jazz bar and the writer chain-smoking at his typewriter are stereotypical images *and* historically accurate. It may be that the world-wide, often rabid and Puritanical, crusade against smoking may eventually reduce the number of smokers to a level that makes this particular stimulant of little importance.

.

---

* These coffee houses were reputed to be places where one could get more than coffee from the female servers.

Finally, there is 'speed'—the amphetamine group. Short-term effects include alertness, euphoria, increased confidence and increased concentration. What more could any creative person want?

Hell, what more could *any* person want? Consider the long-haul truck driver popping these "uppers" to stay awake and alert. Consider the Fifties housewife taking the then commonly prescribed 'diet pills' to slim down and perk up. We drivers would like those truckers to be on the ball. We husbands would like our wives to be slim and energetic. But there is always a rub. The amphetamines are highly addictive, become easily habituated or "tolerated" (requiring increased dosages for continued effectiveness), and once again the over-use and increased potency has truly horrific effects. We don't want those truckers in the on-coming lane to be hallucinating. We don't want our sweet housewives to have schizophrenic episodes before (or even after) dinner.

.

But the creative will always want some stimulant—and easily become addicted to it. It wasn't just Freud and Sherlock Holmes and Erdös that turned to the more potent cocaine as 'better' than coffee and nicotine. Again potency is the issue. One can only expect the future to bring more potent stimulants, and this is likely to be a mixed blessing.

.

There is an interesting recent development in psychopharmacology that, for the more optimistic prognosticators, may seem to offer all the advantages of stimulation, without any notable disadvantages. A new drug called *modafinil* has received a lot of press recently. The military has shown especial interest in it for keeping their troops awake and alert, even under conditions of severe sleep deprivation. It has been reported that the French government's Foreign Legion used Modafinil during covert operations inside Iraq during the first Gulf war, and reputedly the U.S. government is using it now with some of their more elite troops in Afghanistan and Iraq. It has been hailed as more potent than caffeine and far less dangerous than amphetamine as a stimulant. Research on it and similar drugs is being frequently reported in the scientific literature. It is yet to be seen whether or not advances in this area of research will lead to a pill that allows us to function at peak performance without our 8 hours of shut-eye supplemented by coffee and nicotine. I remain sceptical, but perhaps this may well be a new wondrous territory where there be fewer dragons.

.

Therapeutic. *One must be sane to create.* Perhaps not entirely sane, but sane enough. The belief that great artists (and scientists to a lesser degree) are all mad, and madness is somehow prerequisite to genius is nonsense, but that some degree of what many consider 'mental illness' may actually be covariant with or even an important component of creativity may contain a grain of truth —as previously discussed.

.

Alcohol has certainly the most self-prescribed drug to deal with mental and emotional problems. It may even be the least dangerous of therapies. As Tom Waits famously remarked, "I'd rather have a bottle in front of me than a frontal lobotomy." But we live in an age of magic pills, wonder drugs that have done what alcohol couldn't do: release the victims of psychosis from their psychological prisons.

.

But these drugs have serious side-effects, and among the most worrisome is the loss of intensity of experience. Will future developments in psychopharmacology allow cure of a disorder without simultaneously destroying the creative component that is inextricably linked to the disorder? Artists and writers once feared that while psychoanalysis might 'cure' them of their neuroses, it would simultaneously destroy their creativity as part of the 'cure'. This we can now say with confidence is a groundless fear, but it may not be with some pharmacological 'cures'.

.

I remember an interesting radio interview with a sufferer from Turrette's Syndrome. This is a rare neurological disorder that is characterized by uncontrollable twitching and the involuntary utterance of expletives amidst normal speech. (I recently learned a new word in a Scrabble game with my daughter: it is "infix", like suffix and prefix—but *inside* a word. A good example of this is the exclamation a fellow I knew would utter if he sank a hard shot in pool: "beautee-fucking-ful". My friend was a normal guy, but such an exclamation is very Turrette.) What was particularly interesting about the interview with the Turrette victim, however, did not have to do with the verbal symptoms of the disorder, but rather the involuntary muscle twitching. The man was a part-time jazz drummer who was taking a new and entirely effective drug for his condition, but he deliberately went off the drug on weekends, because his jazz band's gigs were on the weekend—and he couldn't play drums, couldn't keep a complicated rhythm, when taking the medication that stopped his involuntary twitching. Sometime before I had read about a practising neurosurgeon who had untreated Turrette's syndrome and was reminded of a cartoon depiction of a brain surgeon exclaiming in the middle of an operation: "Oops, there goes the music lessons!" But this real-life neurosurgeon was entirely expert and never twitched when he was actually operating, just as presumably the jazz drummer never did when he was executing a complicated solo riff. The paradox is fascinating, and recent research has been casting some light on it, but the point is that a drug cure for some disorder might actually disable the recipient's creative ability. We now have surprisingly effective drugs to treat, for example, bipolar disorder.* But if such drugs had been available in the past and were given to Byron, would it have blocked his creative output? Is it possible that for the artist the cure could be worse than the disease?†

.

The answer to the first question is only to be found in deeper understanding of the relationship of dysfunctional personal characteristics to the socially valued side-effects of creative accomplishment. The answer may be disturbing, but evolution isn't concerned with our feelings. The verdict isn't in yet.

.

The answer to the latter question is dependent on how much of a personal sacrifice creative individuals are willing to make to preserve their creativity. One gets a hint of the answer to this question in the silly fear, mentioned above, that so many writers exhibited in the heyday of psychoanalysis: they were afraid that undergoing psychoanalysis would cure them of their crippling neuroticism—but simultaneously deprive them of their inspiration. Most chose in favour of protecting their Muse, even if it was at the expense of their own personal lives. Only when psychoanalysis started billing itself as a pathway to the subconscious mind wherein allegedly lay wonderful treasures to be mined, did artists turn to this pseudo-scientific 'therapy'.

.

Some things are certain. Human beings will always be tempted by whatever frees them from inhibition, by whatever stimulates them, by whatever releases them from mental distress. For this reason the "War on Drugs" cannot ever be won. Besides, the numbers of the 'enemy' are continually increasing, in research labs, legal and illegal, all around the world.

---

*Lithium …. And consider Chlorpromazine and *One Flew Over The Cuckoos Nest*.

† See the comments on treatment of bipolar disorder in the chapter on madness.

CASE STUDIES: ARTHUR RIMBAUD, TIMOTHY LEARY

You wouldn't have wanted your daughter (granddaughter, great granddaughter, even your annoying neighbour's daughter) to marry the poet Arthur Rimbaud or the would-be scientist Timothy Leary. In fact, no sane parent would want any of their offspring to have anything to do with these two reprobates and druggies. They were not very nice people. Nonetheless, they both had a lasting effect on the arts *and* on culture in general. It was an effect inextricably linked to mind-altering drugs and the disordering of the senses. Rimbaud's reputation rests on real creative accomplishment in the domain of poetry. Leary's reputation rests on the effect his charismatic self-promotion and proselytizing of the radical values and attitudes had on the many he seduced with his extravagant claims for the power of hallucinogenic substances. Leary's creativity—and I apply the term reluctantly—lay in the strange domain of celebrity and social influence. So even if he was not really an important creative artist or thinker or scientist himself, he had an even greater impact on a whole (very large) generation than his much more talented precursor Rimbaud. And his influence on some of his truly creative contemporaries is important and useful in understanding and extrapolating from current trends.

.

Arthur Rimbaud is the archetypal *enfant terrible*. It is extremely common for the creative to have been troublesome youths. The 'disagreeableness' trait so requisite to creativity is difficult to abide in the young—who are expected to 'respect their elders' and their values, no matter how absurd those values. The intelligent, creative youngster is likely to think that it is stupid to 'respect' one's elders simply on the basis of their age—as if mere survival to a certain age were more important than the questionable validity of these old fogies' values and ideas. And, of course, they have a point. But the young have a harder time getting away with making this point than those with a few years and accomplishments under their belts.

.

Jean Nicholas Arthur Rimbaud was born on October 20, 1854 in the small town of Charleville, located in North-eastern France. His father was in the military, and abandoned his family when Rimbaud was six. His mother, by all accounts, crossed the boundary from strict parenting into what amounted to child abuse. Yves Bonnefoy's book, *Rimbaud*, suggests that his mother's sadism was fuelled (as is only too often the case) by an insane religious fanaticism. This is confirmed by

her correspondence about death and retribution and by the macabre anecdote about her (at age 75) having a grave-digger lower her into the grave she was later to share with her dead children (including Arthur), so she could have some sense of the eternal darkness to come.
.

It isn't entirely surprising that being raised by such an emotionally disturbed 'single mum' would lead to extreme reaction from a gifted young man. Rimbaud initially displaced his creative energy into academic performance, and his revolt was not, as is often the case, against his teachers and the educational system. Still his academic excellence—which included numerous academic awards and compositions of original verses in Latin—wasn't without some conflict with authority. And upon the publication of his first poem (at age sixteen!), Rimbaud, abandoned his formal education and ran away from home. It wasn't the first time, but it was the last time and for good.
.

He became what sounds like a role model for many of the rebellious youth of the sixties. He grew his hair long, wore shabby clothes, and revelled in shocking the bourgeoisie. He joined the Paris Commune, which was a socialist/anarchist revolutionary political organization of brief duration that had been formed after the Franco-Prussian War had resulted in defeat and humiliation for France. Surely this all sounds familiar: the long hair, the mocking and shocking of the middle-class, the commune and anarchist ideals. But like so many of the run-aways in the hippie era, his fate was not one with a happy ending. It is rumoured that he was raped during his brief involvement with the Paris Commune. Modern hallucinogenic drugs weren't available, but he used alcohol to excess, including that allegedly psychoactive 'green devil's absinthe'.
.

It was at this time that he wrote to his literary mentor Georges Izambard about his poetic—even visionary or religious—agenda: to achieve transcendence from mundane reality by "derangement of all the senses". Rimbaud was only seventeen years of age! It was almost exactly a century later that such "derangement of the senses" became the goal of a whole generation wishing to achieve transcendence, albeit this time along the fast track of ingestion of hallucinogenic drugs.
.

Rimbaud had sent the eminent symbolist poet, Paul Verlaine, some poems, and Verlaine was sufficiently impressed to invite the younger poet to his home. Verlaine had been married to a young woman for little over a year at this time, but his emotions were extremely labile *and* he was bisexual. He immediately fell in love with Rimbaud—who was physically attractive despite his shabby clothes and sullen demeanour—and began a very stormy affair with this brilliant adolescent ten years his junior. The two supplemented absinthe and sex with hashish in an effort to further derange their senses. Soon they became infamous in the Parisian cafes for their outrageous behaviour. Rimbaud once spread his own feces on a café table and 'painted' them with his hand—allegedly to mock painting as two-dimensional and worthless. Verlaine's literary reputation was already well established, but at this time Rimbaud was also making his mark on the literary scene with the publication of *The Drunken Boat* (*Le Bateau Ivre*). So the deadly deranged duo weren't regarded as mere irresponsible obnoxious drunks and druggies, but rather as celebrities whose misbehaviour was thus of interest—and perhaps somehow justified as a means to the higher end of creative accomplishment.

.

This has now become a common justification for misbehaviour. Of course immoral means do not justify any end, even if the end is creative accomplishment. But now—with Rimbaud as unacknowledged 'role model'—many people accept and forgive deranged, irresponsible, and even downright immoral behaviour in artists, behaviour that they would condemn in others, just because they believe that such behaviour feeds creativity. This idea has generalized far beyond the mere derangement of the senses as a path to new insights.

.

For example, the very popular Stanislavski System of Acting emphasizes the need of the actor to relive the emotions of his character, and this has been interpreted as meaning the good actor must have had those emotions, and if he hasn't, well then he must live so as to have them: derange not just his senses but his life. Of course this means that actors taking this instruction literally must behave badly to accurately portray the emotions of a character who behaves badly. I don't know of any actor who assigned the role of a murderer has murdered someone to 'get the feel of it', but numerous actors have sought out experiences on the dark side to prepare for playing the role of someone living there—and made the news for their exploits. Or even, like James Dean, pay for it with his life.

.

This special pleading for the creative has generalized to celebrity status period, whether or not the celebrity is really creative. Some say that we hold our public figures to higher standards than we do our peers. Perhaps this is true sometimes, but more often it seems we are more willing to forgive, even admire, the celebrity figure (no matter how trivial his or her claim to celebrity is) for behaviour we would condemn in our neighbour.

.

I can't say whether the behaviour of Rimbaud and Verlaine was more often forgiven, admired, or condemned by their contemporaries. However, certainly at least Verlaine's behaviour deserved serious censure. Verlaine was at this time physically abusive to his young wife and even their infant son. So it was probably to their relief that in September of 1872 he abandoned them to travel to London with Rimbaud.

.

In July of the following year Rimbaud wanted to return to Paris against Verlaine's desires, and the older poet's tendency to rage manifested itself again, this time against his young lover. Verlaine pulled a pistol and shot at Rimbaud twice. Fortunately Verlaine's drunken aim wasn't good, and one shot missed entirely and the second only grazed Rimbaud's wrist. Initially Rimbaud simply fled and didn't even file charges, but eventually the fear that Verlaine would hunt him down led to charges being laid. Verlaine was sentenced to two years in prison, although Rimbaud compassionately protested and wanted to withdraw the charges.

.

The following year Rimbaud published *A Season In Hell* (*Une Saison en Enfer*). One year later, not yet twenty-one years of age he published *Illuminations*. Both books are considered to be masterpieces that have changed the course of poetic history. And then Rimbaud stopped writing!

.

There is some evidence that he had decided to repudiate the wild Bohemian life of his youth. He certainly never after behaved as he had in his youth. He traveled through Europe, enlisted in the Dutch Army (and then deserted), worked as foreman in a quarry in Cyprus and eventually 'settled' in Africa working as a merchant—where he reputedly had several mistresses, for like Verlaine, he was bisexual with a strong libido. The rest of his life after the Verlaine years is far from dull, but pales by comparison to his youth. He died of cancer in 1891, at the age of thirty-seven without having produced a literary work since he was twenty. Had he 'come to his senses' and in doing so lost the inspiration they gave him when they were deranged? It may be a pessimistic thing to say, but so it seems.

And then on the other hand we have Dr. Timothy Leary who never came to his senses, but who lived to the age of seventy-five without contributing anything very significant to art or science, but who certainly 'contributed' greatly to social change.

Timothy Francis Leary was born on October 22, 1920 in Springfield, Massachusetts. He was an only child. His father deserted his wife and child when Timothy was thirteen. Records of his early formal education highlight discipline problems more than exceptional academic excellence.
He received his B.A. from the University of Alabama (after some serious run-ins with the administration), his master's at Washington State and his Ph.D. at the University of California, Berkeley.

His 'discipline problems' continued when he switched roles from student to teacher. His first academic position was as an assistant professor at Berkeley from 1950 to 1955. After his wife committed suicide, he took a position at the Kaiser Family Foundation, and then, in 1959, as a lecturer in psychology at Harvard. One year later he traveled to Mexico where he tried 'magic mushrooms'—and the rest is history.

He and his colleague, Richard Alpert (who later changed his name to Ram Dass) set up the so-called "Harvard Psilocybin Project". They recruited chemist Albert Hofmann of Sandoz Pharmaceuticals to synthesize psilocybin, the psychoactive ingredient in the mushrooms that had turned Leary on down in Mexico, and LSD, which had been discovered decades before. Leary claims in his autobiography[*] to have administered these drugs to at least five hundred professors, grad students, writers, and clergymen. He claims that the majority of his 'subjects' had mystical revelations. Leary and his associates also gave the drug to prisoners and claimed to reduce recidivism rates by doing this. That psilocybin was administered to all these people is substantiated, but the positive outcomes of the experience are not. And by no stretch of the imagination could their 'research' qualify as a rigorous scientific investigation.

But this dangerous and irresponsible project continued for three years, before Leary lost his job at Harvard for failing to show up for

---

[*] *Flashbacks* (1983)

his own classes, and reputedly supplying—or at very least encouraging use of—psychedelic drugs among undergraduate students. However, by now Leary and Alpert (who also had been fired) had considerable notoriety, and had caught the fancy of Peggy, Billy and Tommy Hitchcock, heirs to the Mellon fortune. They set Leary and Alpert up in a mansion in Millbrook, New York, where they could continue their 'experiments' and 'research'. Reports and rumours of the goings-on at this place are legion, and sometimes contradictory. Some claim that it was a 'spiritual' place for serious study of the enlightening effects of psychedelics; others claim it was one crazy ongoing Bacchanalian party.

This went on for five years or so and in 1966 Leary founded something called the League for Spiritual Discovery and tried to obtain status as a religion, thereby getting legal permission to administer LSD and other psychedelics as 'sacraments'. The ploy failed and the group was harassed by the authorities.* But Leary had obtained great fame by this time and was touring campuses promoting his ideas of better living through chemistry. In January of 1967 he spoke at a "Be-In" at a San Francisco park to 30,000 hippies and reputedly there first uttered the famous words "Turn on, tune in, drop out." Something many in his audience took to heart.

The rest of his life is a complex and sometimes sordid melodrama that, when viewed objectively, is an indictment of his character and/or the allegedly enlightening effects of psychedelics—and I suspect it is an 'and'. His ideas became more and more flakey, including an obsession with colonization of space which led him to oppose conservationists and the ecologically concerned. His troubles with the law, including imprisonment on really trivial drug possession charges, gave him martyr status among his admirers, and he cashed in on this big time. The fact is he was given a pardon for his last prison sentence for testifying with immunity in the prosecution of his own friends in the drug culture and others whose political activities angered the Establishment. (LSD apparently had not endowed him with the virtue of loyalty.) He became a Hollywood celebrity and, as his ideas became more deranged, seemed to attract the interest and affection of many talented writers, including the (then) young science-fiction writers William Gibson and Norman Spinrad. He rubbed shoulders with rock and movie stars and lived in opulence.

* Including the infamous Gordon Liddy (of Watergate fame), who later became buddies with Leary and toured a lecture circuit with him.

Along the troubled way of his life one of his daughters, like his first wife, committed suicide. Then in May of 1996, Timothy Leary took his final trip, if not to outer space, to some great beyond.

.

The verdict is in, and Timothy Leary is a key witness. Spiritual—or any form of—enlightenment is not to be found in a chemical solution. It is true that Leary may simply not ever have been a nice person, but certainly his claims for the power of psychedelics to make one a good, enlightened person is belied by his own life. This is not to say, of course, that these drugs don't derange the senses, that desired state for Rimbaud, for that is precisely what they do. It is to say that such derangement in and of itself is not sufficient for any good—and more often results in destruction than creation. Even when it genuinely seems to aid and abet creativity, as seems to have been the case with Rimbaud, it often results in what military spin-doctors euphemistically call extensive 'collateral damage'. Growing up in The Sixties, I personally knew some of the casualties.

.

However, it is a safe bet that this is one territory that will be explored more and more in the future. It is also clear that the dragons lurking there are very real. The secret agent from the science of chemistry is a double agent: sometimes useful, but never to be trusted.

STRIP MINING THE SUBCONSCIOUS

"The subconscious is ceaselessly murmuring, and it is by listening to these murmurs that one hears the truth."
—Gaston Bachelard (*The Poetics of Reverie*)

"It is surprising, given typically excessive human pride, how often we are willing to disavow credit for our accomplishments and credit imaginary creatures such as God or Some Muse or The Subconscious"
—Hippokrites

The Muse has been renamed recently and her name is Her Highness Subconscious. Freud named her, and those who now worship her are legion. But there still are sceptics, atheists and agnostics, who think a phantom by any other name is still a phantom.

.

"How else can you explain..." begins the defence of everything from clairvoyance (How else explain such an odd coincidence?) to a Supreme Being (How else explain the beauty in the world?) Sometimes there are clear alternative explanations and sometimes not. But the lack of a reasonable alternative explanation does not make the prevalent and popular explanation correct.

.

Probably more artists now believe in The Subconscious than ever truly believed in her previous incarnation as The Muse. And how else can they explain their creativity—the strokes of genius, the gifts of phrase, the sublime melodies, that come unbidden from somewhere out beyond their consciousness?

THE SUBCONSCIOUS: SOMETIMES WHAT YOU SEE IS ALL THERE IS

Perhaps the concept most central to Freud's theory of human nature is the idea of there being something beyond—or beneath—consciousness. It is also the most widely accepted feature of his theory. Not many people really believe that all boys lust after their mothers and wish to kill their fathers; you'd be hard put to find a woman who endorses the notion of penis envy; and although many will admit that sex is a more important motivation behind human behaviour than was acknowledged in Freud's time, it is the existence of a subconscious that has achieved almost universal acceptance as scientific fact. But is its existence, like that of electrons, while unobservable directly, nevertheless substantiated by empirical evidence?

.

Remember that Freud was a wannabe scientist. Perhaps if he had been born in our time, he would have pursued a lifelong career in neuroscience. In fact he did begin his career in neurology. He entered Vienna University in 1873 as a medical student with the intention of applying the methods of science to the understanding of brain function. There he did research work on the central nervous system under the guidance of Ernst Wilhelm von Brücke. So involved was he in neurological research that he took several years longer than usual to complete the required courses for his medical degree. Even then he remained at the university as a demonstrator in the physiological lab. But the physical brain was not as accessible to direct observation as it is now—with PET scans and CAT scans and functional MRIs. (And psychopharmacology was almost a hundred years in the future.) So he eventually went to work at the General Hospital of Vienna, specializing in 'nervous diseases', and soon fell under the influence of the French neurologist Jean Charcot who introduced him to the pseudo-science of hypnosis.

.

It is entirely plausible that if more real information were available about the physiology of the brain, and more tools available to investigate this, Freud would not have resorted to the construction of his entirely hypothetical (even imaginary) 'anatomy' of id, ego and superego. But one must remember that at this time phrenology (the idea that bumps on the skull correlated with brain function beneath) was just fading as a serious branch of 'physiological psychology'.

.

Freud's 'anatomy' was tripartite. To grossly oversimplify, the *id* was the primitive brain fuelled by libido, the *superego* that highly evolved part of the brain one might loosely equate with conscience*, and the *ego* that part that resolved conflicts between instant-gratification id and tight-assed superego. Only the ego was associated with consciousness. The id (the engine) and the superego (the brakes) are largely invisible, which is why men aren't aware they lust after their mothers and why women don't understand why they feel guilty about enjoying sex. Enter the subconscious.

The distinction between consciousness and unconsciousness is obvious. If you fall asleep or get severely whacked on the head, you lose consciousness; i.e., you are no longer aware of sensory input or even of your own existence. The subconscious is different. It is there below the surface—responsible for everything from slips of the tongue to one's preference in sexual partners—when you are fully conscious and think you're behaving rationally; and it is there—guiding and shaping your dreams—when each night you embrace sleep.† It is also, according to many popular theories of creativity (including one to be discussed in the next section) busy at work solving scientific problems and shaping works of art while you go about doing mundane tasks.

Anyway, so goes the theory. But where is the supporting evidence?

Do we do things without being consciously aware of why we do them? The answer most of us would give is a resounding yes. But that wasn't true in the wake of The Enlightenment and The Age Of Reason. Even those most critical of Freud have to admit that he made a significant contribution to human understanding by his emphasis on the irrational: it was a much needed antidote to the naïve idea that man was a purely rational animal.

So it seems there really is need for a term to describe motives of which we are not aware. Yes, the excesses that follow from admitting

---

* Not necessarily a *reasonable* conscience, for Freud felt that excessive, unjustified and neurotic guilt originated in an overactive superego that considered normal desires somehow naughty and nasty.

† Sometimes the terms "unconscious" and "subconscious" are used interchangeably, such as when we talk of "unconscious desires", but for clarity I'll use "subconscious" as the term for cognitive and emotional activities occurring below our conscious awareness—whether awake or asleep.

into our intellectual vocabulary the idea of subconscious motives are almost as horrific as the mounting evidence of our irrationality, but admitting to some lack of self-awareness of our motives does not require accepting such patent nonsense as so-called "repressed memories" or "subliminal persuasion" or "hypnosis". But what is germane here is whether this idea of the subconscious serves to explain some of the mysteries of creativity?

.

Why is the myth of the Muse so resonant? Why do some artists and scientists report being "given" unbidden their most brilliant creations? If creativity was a totally rational process, why does it seem so mysterious—even to those most creative?

THE FOUR STAGES ON WHICH THE CREATIVE PLAY IS PERFORMED

A book published back in 1926, *The Art Of Thought*, has profoundly influenced most thinking about the creative process ever since. In this work Graham Wallas suggests there are four basic stages to the creative process: 1) preparation; 2) incubation; 3) insight; and 4) refinement and verification.
.

This is a useful paradigm, but like most such 'stage' theories in psychology it should be understood only as an approximation and gross simplification of a general pattern. I will, however, use it as an outline for describing the creative process, and suggesting how it works similarly and differently in art and science.
.

Preparation. I think it easier to understand this stage by dividing it into three sub-stages. I would suggest: 1) general preparedness; 2) problem invention or problem discovery; 3) specific preparatory work.
.

Before artists or scientists can embark on any creative journey they need to decide where to go, and before they do that, they need to know of possible destinations. All artists and scientists need a preparatory period of learning about their domain before they can ever hope to expand it. This is what I mean by general preparedness. The poet needs to know poetry, have some conception of what has been written, before he can write something new. If a student hands me a sheaf of his poems to evaluate, I always ask him which poets he reads. If (as is too often the case) he can't name any, I hand the poems back to him. The physicist needs to know physics, the current theories, before he can propose a new hypothesis. Einstein needed to deeply understand classical physics before he could overturn it. Once the artist or scientist has explored the familiar terrain of his domain, he can see its boundaries and choose in which direction to venture beyond these borders.
.

It is useful to describe the next preparatory stage in terms of problem definition, although it seems more literally descriptive when talking about science: The scientist is most obviously working at solving a problem. Einstein discovered a problem with the Newtonian description of gravity, and this set him off. And, of course, any of nature's unsolved mysteries are problems. The trick here *in science* is to discover a problem and clearly define it. With artists, the 'problem' is often self-defined and only loosely defined. It may be an expression of an emotion or a complex idea, the description of an internal or external landscape, the integration of function and form in a building or common object, etcetera, etcetera. The point is that the artist more obviously creates, rather than discovers, the problem that motivates the creative act.*

The third sub-stage of preparation is 'worrying the problem'—or 'playing around' with it. Here both the scientist and artist try out different possible solutions. This may be in conversation or in contemplation; in a lab or in a notebook. It may involve researching what others have done or thought. It almost always involves frustration.

Incubation. This is the most controversial of the four stages. It suggests that at some point the unsolved problem is put aside, left to *incubate*, in the belief that the problem egg will eventually and suddenly crack open, and the newborn solution emerge. It is based on an implicit assumption that is open to question: i.e., that we have a subconscious homunculus who will work on the problem while our conscious ego goes on holiday—and will refuse to solve the problem if our conscious ego insists on hanging around and getting in the way. The novelist and essayist Arthur Koestler, examined the question of creativity in his book *The Act Of Creation,* wherein he cites several examples of incubation followed by sudden insight. One is of the mathematical genius Henri Poincaré taking what was for him an unusual holiday from a rigorous work schedule during which he 'discovered' the Fuchsian Function. Another is the famous story of Kekole having a dream of a snake eating its own tail which led him to realizing that the benzene's molecular structure was a ring.

Insight. So the properly incubated egg cracks open. This is the stage where the solution to the problem comes—and allegedly comes in a blaze of blinding light. Insight! Illumination! The light bulb in the cartoon bubble lights up! Archimedes leaps from his bath and cries "Eureka!"†

---

* Some philosophers have argued that scientists too create, rather than discover, problems, but that is a debate for another time.
† Sometimes it is preceded by an 'intimation' that the solution is at hand, like the aura an epileptic feels before a seizure.

Well, this is certainly an over-dramatization of most insights. For one thing, most creative products are not based on a single insight: they are based on numerous insights cobbled together with mundane logical connections. William Shakespeare may have suddenly exclaimed to himself "Ah, yes that is the word I'm looking for!" or "Ah, what a perfect phrase just popped into my head!" Einstein might have suddenly said to himself "My gosh, what difference is there really between free fall in a broken elevator and weightlessness in outer space?!" But we're still a long way from Hamlet's soliloquy or a theory of relativity. Another indisputable fact is that some insights are *not* sudden, but painfully gradual.

I've tried to analyze my own personal experience of sudden insight and the role it has played in my modest creative efforts in several fields. I know introspection is a suspect methodology, but it is precisely because I do know this that I can see how it can be misleading. What I mean by this is that some of my own experiences could lead me to believe in a busy little homunculus hiding out in my subconscious, but upon sceptical reflection I can see alternative explanations.

My most remarkable experiences of sudden insights mysteriously being delivered up to me are in the area of computer programming. Computer programming largely involves finding algorithms to accomplish some task and then translating these algorithms into a language the computer can understand. Like writing a poem, the real work is not in coming up with an idea, but with the editing. A particular sequence of words doesn't work in a draft of the poem, so one 'debugs' it. A sequence of computer instructions doesn't work in a program, so one 'debugs' it. Although both are problem solving, with programming, unlike poetry, there is a clear criterion for successful debugging: the program does exactly what it is supposed to do. It is for this reason that finding the bug and swatting it is usually a more dramatic, discrete event than it is in writing. And I suspect it is for this reason that my most frequent Eureka experiences have been debugging program, not poems.

On innumerable occasions I've have gone to bed in the wee hours of the morning totally stymied in my attempts to solve a computer programming problem—and have popped awake in the morning with the correct solution, as clear as crystal, in my head. I'd run upstairs to my study and test it, and almost without exception the solution would be correct. If anything could convince one of subconscious problem solving, this should. However, there are several other possible explanations which do not require hypothesizing the existence of a subconscious problem-solver.

.

When I went to bed after struggling unsuccessfully with a programming problem, I was exhausted and probably not thinking clearly anymore. When one is tired one tends to fall into a rut, applying the same or similar futile approaches to a problem over and over again, like the proverbial drunk continually looking for his keys under the street lamp (where he can see better), even though he dropped them a few meters away in the dark. This is called perseverance, and it blocks any new ideas trying to getting our attention. Rest attenuates the loud voice of perseverance, and then new ideas have a chance to be heard.

.

This is related to the well-established phenomenon called the 'Reminiscence Effect' described in the saying that we learn to ski in the summer and swim in the winter: a seasonal break from a skill being learned often results in an apparently inexplicable leap in performance upon returning to the task. When we are learning something new, developing some skill, our performance correlates well with our learning *until* we become fatigued. Then we do continue learning, but our performance no longer reflects this learning, so it doesn't appear that learning is continuing. When the fatigue that is depressing performance is removed, the 'invisible' learning that has occurred shows up in performance as this sudden leap in competence.

.

It may well be that when we worry a problem long enough, we do find paths to the solution, but fatigue keeps us from exploring them. When the fatigue is gone, when, for the drunk under the street lamp, the dawn breaks and the alcohol in his system has been metabolized, he instantly spots his keys a few meters from where he was searching the night before.

.

My Eureka experiences in writing are fewer, but it is true that lines of poetry have come to me spontaneously while riding my bike or walking my dogs—lines that I find no reason to change or edit at all, lines that may even be the kernel of a new poem. However, reflecting on this, I realize that the vast majority of the lines of poetry (or anything) that pop into my head are worthless and soon forgotten. I'm probably just exhibiting some critical sense in paying attention to the rare random lines that have poetic resonance.

The famous story of Coleridge's composition of his great poem "Kubla Khan" is often cited as evidence of pure poetic inspiration, a work delivered to Coleridge whole while under the influence of opium, but which had the ending truncated by the inopportune arrival of a visitor from Porlock. Literary scholars however have found earlier drafts of parts of the poem, casting serious doubt on the veracity of Coleridge's version of it leaping into existence fully formed and in no need of revision.

I've done extensive reading of interviews with writers about their creative process, as well as had many personal conversations with writers about this topic. Most, when pressed, deny that inspiration and Eureka experiences should be given credit for their creations. Stephen Spender, a poet of no little accomplishment, maintains that 'attention', not inattention, is the key to being successfully creative.*

Everyone has experienced flashes of insight now and then, which sometimes are mere flickers and sometimes blinding, but almost all of these insights are only starting points to something of actual significance or else solutions to some problem encountered along the way to completing the creation. Even my being delivered a solution to an annoying debugging problem as I wake from sleep is of little significance really, no matter how dramatic it seemed to me at the time. No programmer has ever had a full length and perfect bug-free program presented to him upon getting out of bed after an insightful dream. No complete novel or poem was ever written by the putative subconscious.

The creative process is a conscious one, and the creative product is no pre-packaged and gift-wrapped present from The Unconscious. This is not to say that connections and relationships that catch our attention, sometimes mysteriously and dramatically, aren't important. It isn't to say that inspiration from delivered by someone we might call Mister Subconscious is misguided. But amateurs and bad writers wait for inspiration; the pros get up and just go to work like any other working stiff, putting in their hours typing away and consciously trying to make sense—or sometimes even make something beautiful.

---

* Spender, Stephen (1952) "The making of a poem" in B. Ghiselin (Ed.) *The Creative Process: A Symposium.* Berkeley: University of California Press, 112-125.

Refinement. This is the revision and verification stage. It is also the stage most often disastrously neglected by those with naïve romantic notions about the creative process. Every teacher who has to grade term papers, and every writer who agrees to look at would-be writers' productions, is aware of this. Those who don't write frequently when finally 'inspired' (or pressed into action by a deadline) tend to treat their output as oracular and even resent any suggestion it could use some editing.

The importance and nature of this final stage is different in science and in art—and different in the different arts. And even differs between different artists working in the same field.

In science, the hypothesis or theory has to be subjected to empirical validation. If it can't, it isn't considered science at all.[*] Many a really clever and novel hypothesis or theory has had its life abruptly ended when put to the test. Even if not put to the death by heartless reality, one could say that in science this final stage is never completed. The never-ending, self-correcting nature of science has already been discussed in some detail. There do not seem to be any absolutely final creative products in science, and scientists who develop a theory often spend their whole lives refining it and seeking verification of it.

In art, however, the situation is different. Works are completed. But there the problem is defining 'completion'. When, for example, a poet 'completes' a poem, he is really saying he has fiddled with it, tried to refine it, and can make no more progress. In other words, since any further changes seem to do no good, it is time to *abandon* it. Only the incredibly vain would assume it was really completed in the sense of being perfect. I have a friend who constantly goes back and—to use his word—tinkers with old poems. However, that some writers produce different versions of a work and return to this refinement stage until their death, doesn't mean that at some point the refinement stage of artistic production does not eventually end. Our understanding of the universe will continue to be subjected to revision. *Shakespeare's Hamlet* or *Mozart's Jupiter Symphony* will not.

Wait. What about *Mozart's Jupiter Symphony*? Is not there evidence that Mozart seemed to be inspirationally delivered of complete compositions and so just had to write them down? Does not his

---

[*] This is a current criticism of string theory in physics.

example disprove everything I've just said about the relative unimportance of inspiration and the great importance of refinement. And what about improvisational music such as much jazz? Or action painting? Or, for that matter, any improvisational art? These are valid questions. And the only possible answer is that like all generalizations about creativity there are exceptions, often very important exceptions.

.

Mozart was extremely unusual in his compositional ability, as any composer will tell you. That other great, Beethoven, laboured long and hard and spent a lot of time refining his compositions, as his extant notebooks prove. His genius was no less than Mozart, and his methods are far more typical. It is also an exaggeration to imagine that Mozart didn't refine the ideas that came to him. Enough of his workbooks exist to show that he did, although not nearly as much as Beethoven or most composers. That musical ideas of great complexity can surface with little need or time for revision and refinement is undeniable. Jazz is the obvious example, and one that never ceases to awe me.

.

So let it be admitted that some art forms are improvisational and some artists (even writers) seem to not need to spend much time in conscious refinement. But there is an important caveat here. The successful improvisational artists have all spent a lot of time in a more general refining of their craft. Jazz artists may improvise brilliantly in performance or jam sessions, but they can only do this because of practice and refinement of the skills involved. The same can be said for improvisational speakers or comics or even another poet I know, who swears he never revises his poems—which certainly seem polished to me.

.

Clearly which has the starring role, conscious effort or intuition and inspiration, in a creative production varies from domain to domain and from individual to individual. Generally, however, inspiration gets top-billing for what is often only a bit part.

CAN THE PUMP BE PRIMED FOR CREATIVE FLOW?

Perhaps the most important question regarding future creative endeavours is whether or not individual creativity can actually be increased. I've already discussed the potential and dangers of attempts to do so by chemical means. Now, having briefly reviewed the rough outline of the creative process suggested by Wallas, the reasonable question is whether creativity can be deliberately and consciously increased if we understand how it works. Whatever the relative importance of the four stages or even the validity of them as a generalizable description of the process, no unequivocal answer to this question is implied—just as looking at the lives of creative individuals does not really point to either an affirmative or negative answer.

By now I'm sure it is obvious that I think the evidence for the importance of inherited characteristics and conscious effort are the primary determinants of creativity. However, this is not to deny the importance of experience or spontaneous intuition. And although my scepticism about self-help psychology is considerable, I'm willing to examine the evidence for its efficacy.

One important theorist about creativity is Mihaly Csikszentmihalyi[*]. He is the author of *Flow: The Psychology of Optimal Experience* and *Creativity: Flow and the Psychology of Discovery and Invention*. His concept of creative 'flow' has flowed downhill into popular culture. "Go with the flow!" "Get in the groove!" "Get in the zone!" I don't mean to make light of his serious description of the altered state that people enter when deeply engaged in a creative activity. Many of the characteristics he describes are also common both to profound aesthetic appreciation experiences and to mystical experiences, a topic I've examined empirically[†].

The hard-headed empirical side of my personality finds his description of the creative personality too much of a 'shot-gun"

---

[*] My son who speaks Czech can pronounce this gentleman's name and tried to teach me how to say it. But in discussing him in class, I fall back on Mister C—meaning no disrespect.

[†]Taylor, Shelley, Stange, Ken. (2008). "Relationship Of Personal Cognitive Schemas To The Labeling Of A Profound Emotional Experience As Religious-Mystical Or Aesthetic". *Empirical Studies of the Arts*, **26(1)**, 35-47.

approach, one that fails the crucial scientific test of falsifiability. In one of his books he lists ten paradoxical characteristics of the creative individual, and while all these apparently paradoxical characteristics probably are true of most artists and scientists, they are also probably true to some extent of most people. They remind me too much of astrological readings (or Barnum Psi personality profiles) which say "sometimes you are introverted and sometimes you are extroverted"*.

But what is even more questionable (and not explicitly suggested by his writing) is the implication that anyone can get into the 'flow'—an idea other authors have promoted relentlessly. Type in "flow and creativity" in an Internet search engine, and you'll find such works as *Writing in Flow: Keys to Enhanced Creativity* (by Susan K. Perry) and numerous other works. Like the old 'Right Brain Thinking' books, Csikszentmihalyi's ideas have resulted in the proliferation of self-improvement books that blithely promise creativity to everyone willing to spend $29.95 at their local bookstore.

Now there *are* perfectly good and instructive books on improving one's memory† or improving one's skill in some sport or improving one's critical thinking—or even improving one's divergent thinking. The good books of this sort are based on solid scientific evidence translated into practical advice. However, just as there are sometimes useful vitamin supplements sold at 'health food' stores, there are many worthless 'snake oil remedies' on the shelves as well. And of course the claims made for vitamin supplements are outrageously exaggerated. So, too, are the claims made for improving one's creativity that are in the promotional material for many books or weekend seminars or packaged creativity kits. A reasonable question to ask is how much one's creativity can really be increased by some book or program.

Among the first to propose a method for improving one's creativity was Alex Osborn. He is the fellow who invented the methodology

---

* Other examples include things like "playful but serious" and "stubborn but flexible". C'mon, aren't we all?

† Daniel Schacter's *The Seven Sins of Memory* is excellent and included in the annotated bibliography.

called *brainstorming**, which is a group problem solving technique he claimed doubled creative output. Osborn was an ad man and eventually a senior executive in a major firm he had saved from bankruptcy during the Great Depression. In 1948 be wrote *Your Creative Power,* a book outlining his brainstorming method, which he had been using at his company for several years.

The method is based on getting a group of people together and following four 'guidelines' for interaction and idea generation.
- No criticism is permitted
- Wild ideas are to be welcomed
- Quantity is to be aimed for without immediate concern for quality
- One idea should serve as a stepping-stone to other ideas.

The principles behind these ideas are sound, and the first one has long been acknowledged as important for the flow of creative ideas. Creative generation of ideas and critical evaluation of them block each other, even though both are absolutely essential; and continually switching back and forth between the two while working is difficult. So it is better to separate them by time.

(A common cause of writer's block is this immediate critical examination of what has been written. Better to write any old shit, and go back two days later and go into critical mode. I encourage my students who can't seem to get started on their term papers to lead off with a vicious diatribe against me for assigning papers. Go ahead I tell them, start your papers by cursing me and telling me you wish a thousand fleas will take up residence in my armpits—or whatever other nastiness you can dream up. And then slowly drift into the topic of the paper. They can—and they better, damn it—go back and throw out the material that does not fit the term paper topic during the 'refinement/revision' stage of their creative process. The point is to get the juices flowing.)

This primary principle underlies the other principles of welcoming wild ideas and aiming for quantity over quality. It is far easier to be

---

* Someone told me this term is politically incorrect, because it could give offense to epileptics! Brain-showering was suggested as an alternative. I suspect my leg was being pulled, but if true I must express perverse admiration for the creativity of these self-appointed language police in finding something offensive in almost any common expression.

critical than generative. One can always throw out the useless stuff generated. It is far harder to start generating anything at all. It should be easier to discard than create.

.

The final principle of using one idea as a stepping-stone to another is commonsensical. If you have managed to come up with something, expand on it. The follow-up ideas may be far better than the original.

.

So Osborn was on to something, although his four guidelines weren't entirely original. But he was dead wrong in an important way: he promoted this as a group problem-solving technique. And a group by its very nature works against all of the principles!

.

I have for years used Osborn's brainstorming as the basis for a simple experiment conducted in the labs for my Intro Psychology course, partially because it disproves the widely accepted belief in its efficacy. Brainstorming was uncritically embraced by organizations that were already biased in favour of group work; i.e., the formal education system and big business. In places where the individuality and disagreeableness of creative individuals has always been suspect and a source of problems with discipline and conformity, a technique that claimed to increase creativity without the social disruption that usually accompanies it was bound to have great appeal.

.

The experiment I have my students do is a simple two independent group design. One group works as a real group, while the other group of the same size (usually four or five people) work individually. Both are instructed in the four guidelines Osborn outlines, although obviously for the group of individuals working without communication the no-criticism rule simply means no self-criticism, while for the real, interacting group it includes not criticizing others' ideas as well. The task assigned is the Unusual Uses task described earlier in the section dealing with CQ testing. Each group, the nominal and real, both have five minutes to come up with as many unusual uses for some common object as they can. Four trials are run with different common objects, and the objects are the same for both groups. In both groups, individuals write down the ideas they themselves come up with, although in the 'real' group they must express these ideas aloud. Duplicate ideas are inevitable in the nominal group where the members aren't communicating with each other, and so all duplicate ideas are only counted once.

.

Even after participating in the experiment the subjects of both groups believe the real brainstorming group produced more ideas. The actual data tell a very different story. Quite consistently about fifty percent more unduplicated ideas are generated by the group working individually. These findings are entirely consistent with numerous other published studies about the efficacy of group problem solving using brainstorming.

.

The reasons for both the technique's failure and the belief it works seem obvious upon consideration. The method fails because it is far easier to not be critical of oneself than it is to not be critical of someone else, and furthermore the fear of criticism is as powerful an inhibitor as actual expressed criticism. The wildest and most outrageous ideas consistently are generated by the individuals working alone. Try to imagine the brazen strength of character required of a student who would aloud suggest using a paperclip as an IUD, a suggestion that has appeared twice now in the data from the nominal group. The uses generated by members of the interacting group are consistently more conservative and unlikely to offend or 'gross out' anyone. Quantity is easier if you're not interrupted by having to listen to someone else's ideas. Increasing the group size would consistently decrease the number of ideas expressed by the real group, as anyone who has ever attended a committee meeting with a large number of members knows: one just sits there waiting to have one's say; one isn't contributing.

.

The belief that the group brainstorming works is probably due to the fact that groups have more fun. If the group is doing at all well at following Osborn's ground-rules there will be laughter and pleasant social interaction. If it feels good, it must be working, right? Meanwhile, the individuals working alone are just sitting by themselves engaged in what they might think is a meaningless exercise. (This may in fact be what inspires them to wild ideas, just to amuse themselves.)

.

Well, so much for brainstorming. Another methodology, one far more elaborated than Osborn's, is that of Edward de Bono. It is called Lateral Thinking and is based on the following metaphor. When digging deep for a solution without success we start to show perseverance and just dig deeper and deeper in the same place, but like someone digging for water, it sometimes makes more sense to stop continuing in unsuccessful vertical descent and, instead, move laterally, step sideways, and try digging in a new spot.

Edward de Bono's various techniques for getting people to 'think laterally' are ingenious and have some demonstrated effectiveness. Moreover, his own creative thinking, especially early in his life, offers support for his ideas. However, like Osborn, he tends to emphasize group creativity—which may be appealing to many normally uncreative people, but really is of less value in the more important realm of individual creative accomplishment in art or science. Edward de Bono is an interesting person and one of the subjects of the next case study, so I'll say no more at this point.

Aside from such adult self-improvement methodologies as proposed by Osborn and de Bono, the ever increasing interest in nurturing creativity in the young needs to be considered in any attempt to predict the future. However, the ways that creativity allegedly is being encouraged and developed in the young also is of questionable effectiveness. In the school system, encouraging creativity seems nothing more than lip-service, belied by actual practice. In the home, it seems to often work against its intended goal by actually decreasing the leisure and personal freedom that is essential to creative development.

It may sound cynical, but it seems to me that the formal educational system has always embraced and then distorted even the most reasonable ideas for encouraging creative or intellectual development. From the good ideas and ideals of the philosopher John Dewey through the insights into cognitive development of Piaget, to the more dubious ideas of the "whole language" theory of teaching reading*, educationalists have often done more harm than good in trying to apply various theorists' ideas in the real world of group instruction.

As 'mere' imparting of information and skills was relegated to only one part of the public educators' responsibility, and as teachers were assigned—or assigned to themselves—such other responsibilities as moral instruction and creating 'team players or 'good citizens', it isn't surprising that the nurturing of creativity would eventually also be

---

* The ineffective and silly application of 'whole language learning' was given pseudo-respectability by the educationist Ken Goodman in the late sixties with his publication of *The Psycholinguistic Nature of the Reading Process*. Goodman tries to use Noam Chomsky's entirely plausible theory of child language development as justification for a guessing and immersion approach to teaching the basic decoding of symbol into sound that is the real basis for reading.

considered part of their mandate. This is expecting a lot of a teacher, especially a teacher teaching a group of diverse individuals. Even just imparting of information to a group of individuals with different aptitudes and prior knowledge and experience is a major challenge that only a relatively small percentage of teachers can meet. It is outrageous to expect a teacher also to shape the moral character of all his students, act as a 'role model', a surrogate parent—*and* teach creativity. What results from these extra expectations being made of a teacher is less time devoted to their fundamental responsibility of imparting information.

.

Individual instruction is a very different thing from group instruction. The old-fashioned apprentice and mentor system is unquestionably superior to the current system. Especially in terms of encouraging those with creative potential, the historical evidence is overwhelming. Very few of the eminently creative liked or were successful within the formal education system. On the other hand, Howard Gardner in his book, *Creating Minds: An Anatomy of Creativity Seen Through the Lives of Freud, Einstein, Picasso, Stravinsky, Eliot, Graham, and Gandhi,* considers a mentor a common denominator in the lives of important artists and scientists. In a one-on-one relationship, a master or mentor may incidentally act as a role model and inspiration to his student, but the crucial thing here is the individual relationship and the flexibility it allows. It is also important that the master is not committed to 'improving society', but is more likely a rebel himself.

.

Jules Henry, in *Culture Against Man*, writes that

> The function of education has never been to free the mind and spirit of man, but to bind them; and to the end that the mind and spirit of his children should never escape Homo sapiens has employed praise, ridicule, admonition, accusation, mutilation, and even torture to chain them to the culture pattern . . . for where every man is unique there is no society, and where there is no society there can be no man. Contemporary American educators think they want creative children, yet it is an open question as to what they expect these children to create. And certainly the classrooms -- from kindergarten to graduate school -- in which they expect it to happen are not crucibles of creative activity and thought. It stands to reason that were young people truly creative the culture would fall apart, for originality, by definition, is different from what is given, and what is given is the culture itself. From the endless, pathetic, "creative hours" of kindergarten to the most abstruse problems in sociology and anthropology, the function of education is to prevent the truly creative intellect from getting out of hand

The point is well taken: the public education system is blatantly hypocritical when it claims to want to create creative students. Creativity is in opposition to the status quo and all public group education is intended to promote and support the status quo. Who do the educationists think they are kidding?

On purely pragmatic grounds, group education requires discipline and orderliness, and, as discussed earlier, the intelligent and creative student usually is a big pain in teacher's ass. A room full of such students is the average teacher's nightmare—and not an environment conducive to orderly imparting of knowledge. The so-called 'open-concept classroom' fad in education of a few decades ago, which was supposed to encourage creativity, has been abandoned, and for very good reasons. It made the classroom discipline required for organized, systematic instruction impossible, so the basic canonical knowledge wasn't being delivered. The majority of students following their own noses in 'independent' studies didn't go very far or deep in these studies without any individual guidance. Even those that did do so would probably have done so on their own initiative—and so just missed out on education in any kind of standard curriculum that could have given them grounding for their creative endeavours. Furthermore, given that creative individuals tend to be introverted and prefer to pursue their interests without disruption by others less intent and serious, even these naturally gifted students wouldn't have found this 'open' classroom any less oppressive than the traditional assigned-seat and listen to the teacher system of the traditional classroom.

I am fortunate in having had a good proportion of students in my elective course on the Psychology of Art with creative potential. Some of them have experienced various institutionalized attempts to teach and encourage creativity. They are almost unanimous in having found it laughable and annoying.

This is not to say that all such experiments in education are futile and pointless and that we should go back to some rigid disciplinary system or not attempt to accommodate the potentially creative student. I once visited, as part of a "Writers In The Schools" program, a publicly funded secondary school that billed itself as a "a school of *experiential* education." This school did emphasize individual student projects, as opposed to some standard curriculum, and my brief experience with the students there didn't indicate they had more lacunae in their education than the typical secondary school student. What it did do was astound me with the depth of engagement and knowledge they had with their chosen projects—and learning in general and respect for creative accomplishment. I talked with one student, and read part of her paper on Dostoevsky, who showed more insight into that great writer than any academic scholar I'd ever read. I believe she was sixteen years of age. There was no dress code at this school, and many of the students would have been immediately called into the Vice-Principal's office at any traditional Canadian school for their weird hair, tore jeans, and t-shirts emblazoned with rude comments. The teachers were just as laid back and obviously just as exceptional as their students. Everything seemed very *laisser-faire* and unstructured, yet there clearly was a whole helluva lotta learning going on. It was almost enough to renew one's faith in group education.

.

However, these were neither typical students nor typical teachers. Both had been selected by an extensive screening process. The screening process for the students was not based on previous academic accomplishment, although some minimum demonstrated competence was required. It was based on anecdotal evidence from parents and teachers that these students didn't fit in with the conventional public education system and seemed to have some intrinsic motivation and aptitude to learn and create. As for the teachers, one to whom I spoke told me how competitive was the application process for being hired at this school. He also said how teaching totally consumed his life—as a true vocation should—and how that was the ultimate criterion that was applied in hiring.

.

So schools like this make a difference, but they are special cases and special cases make bad policy. It may be that all schools can improve in ways that will be less oppressive to creative students, but it is doubtful that, even if they do, it will result in some general surge of creativity, anymore than reading on a book on how to improve one's creativity will.

.

I think it a reasonable extrapolation of current trends to expect more and more creativity-made-easy salesmen to be hawking their wares in the marketplaces of the future, including that lucrative market of public education. They will be selling to men in business suits and to concerned and loving parents and to naïve educationists, but I doubt that this will have much effect on the number of eminently creative individuals contributing to civilization—and probably not even produce a very noticeable increase in creative problem solving in the everyday life of Everyman.

.

However, I do think many other changes in our world will have an effect—a positive effect. The glory that was Pericles's Athens in the fifth century B.C. had causes. (Everything has causes.) But Aristotle and Sophocles, et al weren't the result of books on how to be more creative or of their parents trying to enrich their lives with creative toys or of trends in public education—anymore than The Renaissance or The Enlightenment or the early Twentieth Century supernova of creativity in art and science was a result of such. I've already spilled a lot of words discussing some of the real individual correlates, and possible causes, of creativity, and many historians (far more erudite than me) have devoted their intellectual lives to uncovering the larger social causes. To name but a few: cultural diversity, political upheaval, increasing literacy and easier access to education, collapse in credibility of church or state, 'in-breeding' of the genetically gifted*, even disasters such as The Great Plague, all have acted as catalysts to a surge of creativity. What most of these factors have in common is that they cannot be socially engineered or shouldn't be.

.

If anything primes the pump and increases creative flow, it will be the widespread social and environment changes that have always done so and which are now occurring more and more frequently. The benign change this book focuses on is that of increased communication between scientists and artists.

---

* By this I mean the tendency of the artistic, scientific and intellectual elite to find each other and form sub-cultures where the mated with each other.

CASE STUDIES: EDWARD DE BONO, SALVADOR DALI

De Bono and Dali are two masters of self-promotion who have greatly profited from this skill. Both could be called irrationalists* because of their belief that opening the wellspring of creativity involves rejection of conventional critical thinking. De Bono has accumulated great wealth by claiming to be able to do this for everyone. The infamous surrealist artist Dali's financial success was accomplished through maintaining that he had private and special access to this source.

.

Edward De Bono was born in Malta in 1933. He earned a medical degree from the Royal University of Malta and then studied at Oxford as a Rhodes Scholar, where he received an honours degree in psychology and physiology and a D. Phil in medicine. He then proceeded to complete a PhD, at Cambridge. This was followed by faculty positions at Oxford and Harvard. Such academic credentials are impressive and gave his extravagant claims to be able to make anyone more creative a lot of respectability.

.

In 1967 he published *The Use of Lateral Thinking*, which was a best-seller and did much to revive interest in creativity, as well as offer some solid insights into the nature of innovative thinking. De Bono quickly capitalized on the success of his book by, in 1969, founding the Cognitive Research Trust (CoRT) corporation and selling his get-creative-quick ideas in books—and in pricey seminars that the 'suits' in upper management business and government were only too eager to pay for as a worthwhile investment in future profits. In the following years, he was nothing if not prolific, publishing 75 books which have been translated into 37 languages. It isn't surprising that he has been accused of repeating himself, since his basic ideas are neither many nor complex.

.

But like Osborn before him, although even more enthusiastically than Osborn, De Bono was embraced by the educationists and the corporative executives. What educator wouldn't want to enhance their students' creativity—as long as it didn't interfere with classroom decorum and respect for authority? And what CEO wouldn't want to enhance his managers' creativity, and thus company profit—as long

---

* De Bono, not to his credit, has actually had the gall to criticize Socrates and Plato and Aristotle for being too critical; the implication being not 'creative'!

as it didn't interfere with employee decorum and respect for the bosses?

.

Besides the catch phrase "lateral thinking", De Bono is to blame for the horrible expression so loved by politicians and businessmen— "thinking outside the box". What is his actual contribution to important creative accomplishment? There is no way of really making this evaluation. If one searches De Bono on the Internet, one is flooded with gushing and uncritical testimonials and hundreds of his own self-promotional sites—not to mention sales pitches for his books and seminars. Even unbiased biographical information is virtually impossible to find.

.

Central to his method, as it is to Osborn's, is the idea of suppressing criticism during the creative process. This is not exactly a new insight, but the recommendation of it in the context of educational and business hierarchies is a worthy one. And insofar as it has been followed, it seems to have had positive results. However, it seems highly unlikely that he served as inspiration or mentor to any eminently creative individual. I've certainly never come across any endorsement of his method by any notable scientist or artist. His ideas, like Osborn's, have some validity and practical applications, so he almost certainly has effected some positive changes in both the educational and business sectors. No doubt such watered-down, self-help approaches to creative thinking will continue into the future. How much of a real effect it will have on significant creativity is dubious.

.

The other master of self-promotion (which is certainly a major trend in the artistic community that has increased and will continue to increase) is Salvador Dali. If you ask the average person to name some surrealist artists, those who can name any at all will probably only name one, and that one will be the infamous Salvador Dali. Without denigrating Dali's importance, the reason for this isn't because his accomplishment is greater than others such as Rene Magritte: it is his celebrity.

.

We live in an Age Of Celebrity. I know historians are lining up to point out that this obsession with 'high profile' people is not a new phenomenon. For hundreds of years Royalty held the limelight—and still do to some extent in Great Britain—but with the revolutions in mass communication that have come with the Twentieth Century, celebrity now seems to be available to everyone. Andy Warhol's famous remark about everyone in the future having fifteen minutes of fame isn't far off the mark. Obviously not everyone will really have their fifteen minutes of fame, but the *possibility* of at least transient fame is now within the reach of everyone in the industrialized and 'digitized' world. The expansion of high speed Internet access—and innovations such as YouTube and personal blogs—give everyone with a computer a chance at instant fame—or infamy.

No doubt the general public will still remain fixated on public "personalities" such as movie stars, rock stars, and the rich and famous—especially if they misbehave. But there is a notable change of focus here. Actors and performers are artists, and the attention they are receiving is not because of having Royal blood or inherited wealth or power.

This is not to say that artists were not often accorded fame and celebrity in the more distant past, but their fame was limited to a much smaller group of admirers. Beethoven or Rimbaud certainly were both *causes célèbres* in their time, but most of their working stiff contemporaries probably wouldn't even recognize their names. It is true that Franz Liszt was the equivalent of a rock star back in 1842 when "Lisztomania" swept Europe, just as "Beatlemania" was to sweep America 120 years later. The difference has to do with the numbers of fans, the extent of the recognition and fame. A Liszt concert might be attended by a few hundred people, and those are the only people to have experienced his performance.* The Beatles reached millions with their 1964 performance on the Ed Sullivan TV show. Their records have sold 20 or 30 million copies.

It is said that everyone wants to be rich and famous, and that simply isn't true. Most people want to be financially well-off enough to indulge their desires but really wouldn't want the hassle great wealth brings with it. And many people would really prefer anonymity to fame. The hassles associated with fame are only too well established. Of those that do wish for fame, there are two types: one type is the person who naively wants fame for its own sake, and the other is the person who wants fame as a means to an end, be that end wealth or adulation or exposure of his or her creative products. Most writers would like to have many people read their books. Fame is the means

---

* And flipped out over his histrionics. Women in a frenzy, that makes a Beatle concert look like a performance by the Church Choir, apparently threw their undergarments on stage.

to that end. Since introversion is common to writers, it isn't surprising that many are very uncomfortable with celebrity and fame and see it as only a means to an end. The reclusive J.D. Salinger is but an extreme example.

One has to wonder how many contemporary artists are more concerned with fame for its own sake or just as a means to ends such as wealth or attracting attractive sexual partners. And how many primarily see fame as a means to the more noble end of having their work noticed and appreciated—with, but of course, any side benefits most gratefully accepted and enjoyed, yet still secondary.

Where to place Dali? He is far too serious, productive* and significant an artist to dismiss as a mere publicity seeker. No doubt his personality was not that of a self-effacing introvert, and his motives for his brilliant self-promotion campaign may be less than pure†, but certainly his central motive was to get attention to his work, not to himself, or he wouldn't have worked so hard.

Salvador Domingo Felipe Jacinto Dalí i Domènech was born on May 11, 1904 in Catalonia, Spain. Dali always claimed to have Arab lineage, but there is no evidence of this. His father was a lawyer and extremely strict and distant, so the young Dali became extremely attached to his mother, who died when he was sixteen years old—an event he always reported as traumatic.

Dali's father did appreciate his son's talent and had already organized an exhibition of his drawings in the family home a year before his mother's death. At age eighteen Dali began studies at the San Fernando School of Fine Arts in Madrid. He was already displaying his histrionic nature by dressing in a way more common a century before. He was also already displaying his famous arrogance, for he was expelled from the academy before his final exams for claiming that no faculty member at the institution was sufficiently competent to judge him.

In 1928 Dali went to Paris where he befriended Picasso and Miro and became part of the surrealist movement started by Andre Breton. In

---

* 1500 complex and technically brilliant and precise paintings!

†He was notorious greedy. The surrealist writer, Andre Breton, nicknamed him "Avida Dollars", which is an anagram for Salvador Dali.

1929, at the age of twenty-five, Dali met and wooed his muse and future wife* Gala (Helena Dmitrievna Deluvina Diakonova) who was a decade his senior and married at the time to the surrealist poet Paul Eluard. The same year he collaborated with Luis Buñuel on the short surrealist film classic *Un Chien Andalou*.†

.

Largely because of his gift for self-promotion and flamboyant personality, he drew more attention to himself (and his work) than most of the other surrealists. It was all fame and glory from his late twenties until his death in 1989 at the age of 84. It wasn't all a bed of roses however. When his beloved and indefatigable soul-mate Gala died before him, he became profoundly depressed and attempted suicide.‡

.

Salvador Dali is a biographer's—and tabloid writer's—delight, for his life is more packed full of bizarre, wild, and crazy events than seems possible—and he was very public about it all. However, it is inappropriate—and would take volumes—to try to deal with them here. What is germane is the role self-promotion and publicity have recently played—and most importantly will continue to play—in inspiring creative endeavour and having the results of these endeavours reach a larger audience.

.

Dali was a self-proclaimed genius of unique stature and would be the last to claim everyone can be creative by simply following in his footsteps or using his inspirational technique.§ His influence has not been one of teacher of creativity, but rather as a 'role model' for creativity. His histrionic and flamboyant persona has been widely

---

* Gala divorced from her husband in 1932 and in 1934 Dali and Gala were married in a civil ceremony in Paris.

† Dali's dark side, which was intertwined with his increasing religiosity, led to him denouncing Buñuel in 1942 as an atheist. This resulted in his surrealist colleague and friend being fired from his position at the Museum of Modern Art and subsequently blacklisted by the American film industry.

‡ There are some sources that maintain was not suicide, but rather an accident. What is undeniable is his profound depression and grief.

§ He claimed to use what he called the Paranoaic-critical method for gaining access to his subconscious, which was his wellspring of artistic creativity. This 'method' is never clearly delineated and is just one more surreal product of his incredible imagination.

imitated by young artists trying to draw attention to themselves and their work. Along with this comes the implicit assumption that extremely unconventional life styles such as his are a direct path to creativity.

.

Of course this isn't true. While experience and openness to experience is essential to creativity, it is a case of 'necessary but not sufficient'. Like taking mind-altering chemicals, there is no evidence that searching out extreme experiences is going to increase creativity in most people. Admittedly it may sometimes add fuel to the fire of those already burning with creativity, but that is a very different thing entirely. De Bono's and Osborn's advice to suppress the critical while trying to be creative can help a writer get past writer's block. And searching out new and unusual experiences can have the same effect. However this is very, very different from suggesting that shutting down one's critical faculties or experimenting with 'extreme' living is going to make one creative.

.

The other way Dali points to future trends, already very evident, is the application of celebrity to drawing attention to creative work. The ever escalating high jinks of rock musicians is an obvious example. But even in the visual arts performance art draws more and more artists—and attention—largely because it puts the artist's extreme behaviour centre stage. We have come a long way from the relative anonymity of medieval artists.

.

The negative side of this trend is the shifting of focus from the art to the artists, and too often these are the less gifted artists. The positive side of this trend is that the tabloid mentality or—to put it more kindly—the fascination with the human being behind a creation can widen the audience for new innovative work. Dali is a prime example of this. Surrealism would not have so quickly reached such a wide audience and been so widely appreciated had he never left his studio. And, just maybe, his work would not been seen as radical and interesting as it is.

THE MAGIC OF JUXTAPOSITION

The tuba is on fire!

.

The subject is being given a free association test. He is a bright young man. He is wearing a suit and a bowler hat.

.

"Table?" "Chair!" "Why?" "Both are furniture."

.

"Man?" "Woman!" "Why?" "Both are people."

.

"Door?" "Window!" "Why?" "Both are openings in a house."

.

"Cat?" "Dog!" "Why?" "Both are pets."

.

"Tuba?" "Trumpet!" "Why?" "Both are brass musical instruments."

.

Good answers. Logical. The items do go together, and he can even explain why. But he's not scoring high on creativity. Why? Because the tuba is on fire. Explain that!

.

Rene Magritte is one of the most intriguing of the surrealist artists because he is so various and ingenious in his bizarre visual juxtapositions. One of his paintings is of a tuba on fire. Explain that!

.

He refused to do so.

.

So I will.

PUTTING THINGS IN CONTEXT, STRANGE CONTEXTS

Surrealism points the way to one of the most promising new territories open for aesthetic exploration and exploitation. You want I explain that?

Okay. But first it is necessary to define surrealism. Although it probably is the most popular of the many twentieth century art movements—and the one which has best passed the test of time—probably most people would find it hard to explain what it actually is. Yet surrealist visual art is everywhere. The covers of music CDs and books often feature surrealist artwork; one sees very little cover art that could be considered abstract or fauvist or cubist or even impressionist or expressionist. Salvador Dali is known to millions* and epitomizes surrealism in their minds, even though he repudiated the movement, just as the surrealists repudiated him. "That's surreal!" is not an uncommon exclamation in response to anything out of the ordinary. But what exactly does 'surreal' mean? It doesn't just mean out of the ordinary, unusual or bizarre.

The precursor of surrealism was Dadaism, a nihilistic artistic and intellectual movement that mocked traditional aesthetic, cultural and even ethical values. Dadaism was born in Zurich in the midst of the horrors of World War I, and is often said to have been a reaction to the hypocrisy and complacency of the bourgeoisie, which the Dadaists blamed for the collapse of Europe into violent, nationalistic barbarism. The surrealists shared with the Dadaists a deep contempt for the bourgeoisie. But they interpreted the war as a sign of man's profound irrationality, and, following Freud's lead, they felt that exploring this irrationality was an important enterprise.

Freud's influence on the surrealists cannot be underestimated. They embraced not only his theoretical ideas regarding the subconscious and mankind's unacknowledged animal nature, but also his actual therapeutic methods—which they converted into procedures for the creation of art. Originally the most important of these methods was "free association", where the client on the analyst's couch just says whatever pops into his mind, and thereby lets his subconscious thoughts and passions bubble to the surface.

---

* This is proof of the effectiveness of self-promotion in an age where fame has more to do with the image of the artist than with the artist's images.

Why not do this in writing? And so surrealism began as a literary movement. The term was coined in 1917 by the great French poet Guillaume Apollinaire, but it was Andre Breton who is usually considered the founder of the movement, for it was Breton's *Surrealist Manifesto* of 1924 that initially defined the movement and distinguished it from Dadaism. But well before this publication Breton was practising what he was to preach: he had written (in collaboration with Philippe Soupault) an "automatic book", *Les Champs Magnétiques*. Automatism, as it came to be called, involved 'turning off' one's consciousness, and with it any attempt to direct, control or edit whatever came pouring out of one's pen. The results might have held some interest for psychoanalysts, but are basically unreadable for most people. The technique was attempted by painters as well and led to the *Automatiste* school of visual art, which when its practitioners actually practised what they preached, was just as incapable of holding a percipient's interest as books like *Les Champs Magnétiques*.

However, many visual artists associated with the movement rejected automatism as a creative method while still embracing the goal of probing the subconscious and using its contents as artistic material. They took their inspiration from another psychoanalytic technique for getting at the irrational and the subconscious and the emotionally intense, but repressed, part of the human psyche: dream analysis. Dreams are not logical. They involve incongruent elements thrown together. Time is distorted. People and things and places metamorphose inexplicably into different people, things and places. Yet they have some coherence *and* they often induce intense and vivid emotions. Induce intense and vivid emotions. What artist doesn't want to do that?

Surrealist experiments with free association couldn't do that. Visual artists took as their role model the psychoanalyst's patient's dream, not his totally uncontrolled free association. They pulled in the reins, but not so far as to demand rationality and logical coherence, but far enough as to have some control and create something with some coherence—as would be found in a dream.

And so it is that the most succinct and accurate way to describe surrealist paintings is to say they are "dream-like". This version of surrealism soon became what dominant. Originating in the visual arts, its influence soon flowed back into writing, washing away the old automatistic approach*, and then quickly spread to film and other artistic genres.

.

"Sur-real" means beyond real. However, to be beyond real implies the existence of the real. Dreams are like that. They are not *random* images. They arise from an underlying reality that is transformed, distorted and then mysteriously enriched and brilliantly coloured with emotion.

.

So in the visual arts the dream became the central matter. The works of the great classic surrealist painters all share this magical dream-like quality. Look at almost any René Magritte, Max Ernst, Yves Tanguy, Giorgio De Chirico, Juan Miro, Man Ray, or Salvadore Dali painting, and you'll feel like a voyeur peeking in on someone else's strange and very personal dream.

.

One of the notable characteristics of dreams is incongruity, not abstraction. My wife has dreamt of sidewalks being literally rolled up. This is a metaphor for quiet evening in small towns made literal, a profound incongruity. I have dreamt of dinosaurs grazing in my backyard. This is anachronism, the incongruity of placement in time. Things in dreams get displaced and get juxtaposed in ways that are not natural or even possible in the real world.

.

And it is precisely this that focuses our attention, and in doing so lets us see things afresh, where normally they would be mere background.

.

I would go so far as to maintain that it is this surprising juxtaposition of things that define surrealism. Flames aren't odd. Tubas aren't strange. But flames coming from a tuba are. A picture of a tuba doesn't call our attention to the tuba. We tend to simply identify it—and so file it away without attending to it. But if the tuba has flames leaping from it, our identify-and-file reflex is knocked out of kilter. How can it be a tuba if it is on fire? So we look more closely at the tuba. Our attention is commanded. And then the painterly craft that successfully welds the flames to the tuba somehow transform both flames and tuba into something more than either—and gives each a deeper independent meaning and resonance

.

---

* What is now often called "magic realism" in literature could just as aptly, if not more so, be called surrealism.

Now surprising juxtapositions (which are, of course, the same as surprising connections) is also a defining characteristic of originality. So in some sense, surrealism is itself a metaphor for creativity.

.

Surrealism, as it has evolved, is the connecting link between the objective (the rational and empirical world of science) and the subjective (the irrational and internal world of emotion). It has three characteristics that account both for its popularity and its future potential. First, it is not abstract, for it is usually composed of representations of concrete and recognizable objects and experiences. Two, it is not without some inherent coherence, for while it may not be linear or logical, still there is enough of a narrative element to satisfy that basic aesthetic need for story. Three, it is not inaccessible to those without specific knowledge or refined sensibilities, for it seems to tap universal and fundamental emotional reservoirs.

.

.

DISCOVERING NEW CONGRUENCIES IN INCONGRUITY

Because we tend to think in dichotomies, art and science are not the only things to be mistakenly seen to be in opposition: emotion and reason are also mistakenly considered to be antagonists. We say a good decision is 'rational', and, conversely, we often attribute a bad decision to the person "giving in to emotion". But maybe even more than science and art, reason and emotion are really collaborators, and decisions based purely on reason are as damaged as those based purely on emotion. It is entirely 'rational' to decide to commit evil actions if they benefit you, and you have no danger of being caught. It is emotion, not tempered by the restraints of reason, which leads to many heinous so-called 'crimes of passion'. This much is obvious, but it is more subtle than that.

On September 13, 1848, a usually very conscientious and cautious railway worker named Phineas Gage screwed up big time while setting an explosive charge to clear some rock obstruction. The resulting premature detonation launched into the air a large, iron tamping-rod like a huge, blunt arrow. It went flying through Mr. Gage's head and landed thirty meters away. It took out a large part of his left frontal lobe on its flight. Amazingly, Gage survived and even was able to return to work not too long afterwards. Examination by physicians indicated no deficit in cognitive function: he had not become a moron or a vegetable. However, allegedly his personality profoundly changed after the accident. While formerly quiet, reserved and conscientious, he became, post-trauma, loud and rude and unable to make even the most quotidian decisions that we now call 'life-skills'. He drifted from job to job, and by all accounts consistently messed up his own previously well-ordered life until his death in 1860.*

António Damásio, a neuroscientist and philosopher, is the best known of recent theorists on the relationship between reason and emotion. And he takes the tale of Phineas Gage as early anecdotal

---

* It should be acknowledged that the psychologist Malcolm Macmillan questions the historical validity of this classic, much-cited case, in his book *An Odd Kind of Fame: Stories of Phineas Gage*. No matter, ultimately, for much neurological evidence exists to support that damage to the prefrontal lobes profoundly affects the interactive nature of emotions and reason.

evidence of the importance of the integration of emotion and reason. He uses his own (and others') extensive neurological research as more substantial evidence of this relationship. Damásio's interest in this is not new. Two early competing theories of emotion foreshadow it: they are the James-Lange and Cannon-Bard theories.* The latter suggests we cry because we are sad. The former proposes the counter-intuitive idea that we are sad because we cry! It is an oversimplification, but Damásio is in this camp. Smile and you'll be happy.

.

Without going into the details of the neural circuits involved†, the evidence is mounting that we respond emotionally to a stimulus *before* we cognitively 'label' it. This is a fascinating topic, but what is relevant here is a corollary of this theory: this *ex post facto* cognitive labelling is linked to our physiological, emotional responses, which are 'burned' into the emotional pathways of our brain at the time of the experience. This can be used to explain everything from Post-Traumatic-Stress-Syndrome to acquired phobias and fetishes.

.

And it can explain why emotion, deep emotional memory of emotion, is just as essential to decision making as reason. You are a happily married man deciding whether to succumb to the obviously seductive actions of a lovely woman with whom you work. You did so succumb once before, and the complex and contradictory emotional memories of doing so are essential to making a 'good' decision. You weigh the guilt you felt for the hurt to your beloved wife that your last decision produced against the sensual delights your illicit lover served up to you. These items on the scale are *emotions*. The scale itself may be *reason*, but the items being weighed are emotions.

.

We need emotional memories to make our decisions, even in such a simple case as deciding what to eat for dinner. Did the last time I had a hot curry vindaloo make me *feel* good having it, and did it keep me up all night with heartburn and indigestion? If our memories were not coloured with emotion, how could one possibly make a decision

---

* The James here is the great philosopher and psychologist William James.

† The limbic system is central, which includes the amygdala and the hippocampus, as well as the prefrontal lobes which seem to be as close a decision making homunculus as neurologists have been able to find.

about what to order next time we're at an Indian restaurant? How good *felt* the eating last time? How bad *felt* the aftermath?

.

And what relevance does this have to surrealism? Surrealism implicitly acknowledges the relationship of emotion to reason. It not only acknowledges it, it capitalizes on it to produce a product that is a balance between emotion and reason, between the Dionysian and the Apollonian.

.

Surrealism does not reject rationalism, as is often said. No, surrealism tries to be more real by going beyond conventional reality. And it tries to be more rational by going beyond conventional rationality. The way to go beyond conventional reality and conventional rationality is to tap into emotion. This is what we do when we make what we mislabel as 'rational' decisions: we place on the scale of reason our various weighty emotions.

.

The best surrealist art melts down and merges emotion and reason.* Incongruity is its tool, and it is an effective tool to help the creative find new congruencies. This is why I think it will continue to be the most important influence on future art—a place that creative artists will continue to colonize well into the distant future.

---

* A reasonable symbolic interpretation of Dali's melting clocks.

THE MIRACLE OF METAPHOR—IN SCIENCE

A metaphor is an implied relationship between two very different things, so incongruity is inherent to all metaphors. There are unfortunate people with neurological damage who cannot grasp even the simplest of metaphors and so interpret everything literally. Of course poetry (and surrealism) is lost on them, but so are common sayings such as "the grass is greener on the other side of the hill", which they will adamantly maintain cannot possibly be true and is illogical.

.

Too often scientists are seen as suffering from this disorder, as taking everything literally, when of course nothing could be further from the truth. Juxtaposition, putting the commonplace in a new context, is just as common a tool of scientific investigation as it is of surrealism. The already mentioned and classic example of such juxtaposition that inspired Einstein (i.e., the man in a free-falling elevator and the astronaut in outer space) is as surreal as the cigars floating in the air in a painting by the surrealist Magritte. One can never again perceive free falling objects (or cigars) quite so complacently.

.

It is when something doesn't fit, doesn't make literal sense, that we are forced to look more closely. It is this subsequent close examination of what seems commonplace that inspires both artists and scientists. Neither simply dismisses any such disparities as illogical.

.

Science is surreal. It is just as shocking as the most shocking art—and provokes even more vitriolic and even violent reaction from those shocked. Nowhere is this more obvious now than in the field of biology. The surrealism of relativity theory and quantum mechanics in physics is even more bizarre, but it is too esoteric to get much attention from the general public. Biologists, on the other hand, make juxtapositions that are closer to home, more disturbing to our fragile egos.

.

Position *Homo sapiens* on a cosmological, or even geological, time line and our self-importance is diminished almost to the vanishing point. Or, like the scientist E.O. Wilson, find connections between those pesky ants at a picnic and human social interaction. Or study baboon behaviour and suggest it has some relevance to understanding office politics or closing time at the singles bar. Or watch rats press a lever for food pellets and dare to say they show the same patterns of behaviour as the slot-machine junkie. Illogical! C'mon! We aren't ants, apes, or rats!

.

Often, at least in the Twentieth Century, artists have intentionally tried to shock and disturb. Scientists, on the other hand, have rarely had any such intention. However, it is the scientists that have been most provocative, in spite of their intentions. Science is perceived as being more threatening. Books are banned and art exhibitions closed down and artists subjected to censure, but historically scientists have had it far worse. While less intending to offend, they nonetheless offend us more. Perhaps this is because the offence isn't seen as merely that of a deviant individual or of a small coterie of individuals, but rather as a large community unanimous and united in what they present as real and valid. Perhaps it is because science, unlike art, makes more of a claim to universality of insight into the nature of things.

.

As the diversity of experience and knowledge continues its exponential climb upwards, inevitably more and more unsettling and disorienting juxtapositions and metaphors result. And as these disturbing connections are made, not only art, but science as well, is going to seem 'surreal' and require a greater flexibility of thought and perception than ever before. I expect to see this, but I also expect, based on recent controversies, to see more indignant and angry reaction from those whose understanding of metaphor is limited.

.

Who would have predicted that an entomologist writing a book on ant society could have found these insects a useful metaphor for understanding human society? And who could have predicted that this would have resulted in him being demonized as some kind of neo-Nazi eugenicist? (This man, the eminent scientist E.O. Wilson, is one of the subjects of the following case study.)

.

As artists more and more come to see scientists as their allies, presumably they will join forces with them against the repressive ignorance scientists have to deal with today. Here is where I think—hope, believe—the confluence of art and science will produce a strong enough current to wash aside and ashore those who would try to dam the flow toward the fertile and surreal lands downstream in our future.

CASE STUDIES: RENE MAGRITTE , E. O. WILSON

I've chosen the artist Rene Magritte and the scientist E. O. Wilson as the subjects of this case study because both are very radical in their search for common ground between apparently unrelated, incongruent things, yet both were unassuming and unpretentious despite their radicalism. Both quietly devoted their lives to improving our perception of the world, so we could see and appreciate its wonders more clearly.

.

Rene Magritte has always been my favourite surrealist artist, perhaps because he so effectively combines philosophical (and rational) and artistic (and irrational) components in his work. He is the master of juxtaposition, able to find beautiful congruence in incongruity. His work makes one see everything with fresh and innocent eyes by simply changing the context. So at least for me he is the archetypal surrealist.

.

And again there is a paradox, for he repudiated surrealism early in his career. While living in Paris in the late twenties, Magritte became disillusioned with his fellow surrealists whose obsession with the irrational, and with shocking the bourgeoisie, he found superficial. Apparently he also found the idea of using drugs and dreams as a source of inspiration counter to his own more conservative nature, and when the depression struck in 1930, he returned to his native land, settling in Belgium and working as a commercial artist to make ends meet.

.

Magritte was born in 1898 in the town of Lessines in Belgium. In 1910 his family moved to Châtelet. Two years later his mother drowned herself. No doubt this affected him profoundly, although he has denied this. He became deeply interested in religion, even dressing up as a priest, and during summer holidays in Soginies he made the ancient local cemetery his playground. Many of the images in Magritte's paintings are funereal and have a morbid, haunting quality.

.

At the age of twenty Magritte moved to Brussels to study at the Académie des Beaux-Arts. There he made friends with other painters and writers and had his first exhibition at the Galérie Giroux. In 1922, Magritte bumped into a girl he'd known as a teenager,

Georgette Berger, and they promptly fell in love, marrying soon afterwards.

.

It was in 1927 that Magritte moved to Paris and hung around with Andre Breton, Salvador Dali, Paul Eluard and the other surrealists who were becoming notorious on the Parisian art scene. It was here he had his first one man show. But this bohemian period of his life was short-lived. Three years later he was back in Brussels where his was to spend most of the remainder of his life, working in relative isolation. He died of cancer in 1967 at the age of 69.

Magritte is exemplary of several things which I expect to become more common but which were not that common at the time. One is a concern with technical proficiency and the other is a lifestyle that focused more on creative production than complex personal relationships and ego.

.

Regarding technical proficiency, most art critics would agree that of all the surrealists Dali and De Chirico* and Magritte were the best and most conscientious draughtsmen. Part of the popular appeal of these artists over other surrealists is attributable to this fact, for no one can say (as has been said of Miro, for example) that anyone without any skill at representation could have painted "like that". This is not to say that should be a major criterion—or even any criterion at all—for judging a painting, but it does say something about the priorities and serious commitment of the artist to the final product.

.

This brings us to the second defining characteristic, one definitely *not* shared with Dali: the lifestyle of the artist. Magritte's life was, compared to other artists of the time, very staid and even conservative. He remained married to his childhood sweetheart his entire life, and, while no hermit, did not immerse himself in the turbulent social world typical of early twentieth century artists. His art was just as shocking to the bourgeoisie and conservative artistic community as that of other surrealists, but his life was not. He did not deliberately draw attention to himself, unlike Dali. Instead, Magritte just wanted to draw attention to the *world* by creating works that made the viewer see objects in a new and startling light.

.

---

* De Chirico was an early inspiration to Magritte. Allegedly the twenty year old Magritte was moved to tears upon first viewing the elder artist's "The Song Of Love".

To say he was conscientious is gross understatement. He produced over a thousand paintings in his lifetime, and these are beautifully crafted realist surrealist works, each of which must have taken long periods of time to complete. He undoubtedly had little time left over for social or sexual high jinks. His was a highly focused creativity. His work did the important task he had set for himself: make people see sharply what goes unnoticed in its usual context—and appreciate its mystery and magic. As he once said: "Art evokes the mystery without which the world would not exist."

.

It is superficially paradoxical that Rene Magritte, who as a premier member of an art movement intended to shock and disorient conventional perception of reality, was never really a personal target of those his work outraged, while that other gentle and conciliatory soul Edward Osborne Wilson, who never intended his work to offend anyone, has been demonized. This is a good and distressing example of how those who challenge, no matter how tactfully, leftist establishment viewpoints are just as likely—or even more likely—to be attacked as those who set out deliberately to be radical. One particularly disgusting example of this was when members of the self-titled "International Committee Against Racism" poured a pitcher of water on Wilson's head at a conference, chanting "Wilson, you're all wet!"

.

Unlike Richard Dawkins or James Randi, Wilson is not confrontational in any way.* I recently listened to an interview with him, and if any aspect of his personality stood out it was his warmth and gentle nature. In fact, he recently published a book that is a plea to the religious right and the scientific community to call a truce in the so-called 'science wars' and join forces to fight the more important battle against global environmental disaster. *The Creation: An Appeal to Save Life on Earth* is written in the form of letters to Southern Baptist minister which argues that secular humanists and those who believe in God must put down their weapons, ignore their differences, and concentrate on saving the glory of nature that both appreciate deeply.

.

Wilson has an empathetic understanding of Christian fundamentalists that is rare among scientists, for he is a down-home, southern boy, born in Birmingham Alabama in 1929—where and when there were

---

* Dawkins (and almost certainly Randi) are admirers of Wilson, as are most scientists. It is only the politically contaminated (e.g., Stephen Jay Gould) that demonize him.

virtually no atheists or agnostics to be found. Like Loren Eiseley, he was a child naturalist. However, at age seven he damaged his eye in a fishing accident, and his impaired vision and an inherited hearing disability caused him to focus on the smaller things in nature: insects. It is in this field of entomology that he built his early reputation. At sixteen years of age he was beginning a survey of all the ants in his home state of Alabama. He earned his B.S. and M.S. from the University of Alabama (under public funding, for he was a 'poor southern boy') and went on to get his PhD from Harvard.

His 1975 book, *Sociobiology: The New Synthesis*, catapulted him to fame and controversy. Because his chosen specialization was ants*, notorious for their complex social organization, he saw a congruence between the two apparently unrelated fields of sociology and biology. The book would undoubtedly be considered a classic study of social structure in the insect world but would only have found an audience in one corner of the scientific community were it not for the last chapter in which he stated too baldly what should be obvious: humans are shaped at least as much by genetic inheritance as by culture, and there are limits to what we can do about human nature by changing environmental factors.

The quasi-science of sociology is terribly tainted by leftist politics, and biologists only too aware of the importance of genetic and evolutionary factors, are often misperceived as being on the far right of the political spectrum. So perhaps it was inevitable that Wilson would be targeted for daring to introduce hard scientific evidence of the importance of nature, as opposed to nurture, into the domain of sociology. He was promptly accused of racism and misogyny because of suggesting some human beings are born with better genes than others. One must remember that this was 1975, a time of great (and largely positive) social change and reformation, but also of appalling fascistic hostility to anyone who questioned the dubious philosophical justification of social engineering. He and his really modest and reasonable ideas produced a controversy that made big news, including the cover of *Time* magazine and the front page of *The New York Times*.

---

* He was originally interested in flies, but because of the unavailability of pins during World War II, he switched to studying ants. If pins for pinning flies had been available, it is likely he would never have ended up as a controversial figure in the history of contemporary science.

It wasn't bad enough that he was an atheist and scientist who understood and explained the way natural selection worked (thus raising the ire of the religious right), but he also had the gall to suggest that we are as much a product of our genetic inheritance as any environmental factors (thus infuriating the leftist social engineers). Ironically, Wilson had considerable sympathy for both. He had an almost mystical attitude toward the natural world: "Nature holds the key to our aesthetic, intellectual, cognitive and even spiritual satisfaction." He was a supporter of social change that would benefit both the fittest and the less fit: "It's obvious that the key problem facing humanity in the coming century is how to bring a better quality of life—for 8 billion or more people—without wrecking the environment entirely in the attempt." One is naturally reminded of C.P. Snow, who managed to alienate both the scientific camp and the artistic camp, while really supporting both.

How did Wilson respond to the controversy? Rationally and coolly, but unflinchingly. In 1979 he published *On Human Nature* (which won the Pulitzer Prize) in which he elaborated on the final chapter in *Sociobiology*. In this book he challenges the preconceptions and prejudices that well-meaning individuals use to justify Draconian measures to shape society into the way they think it should be structured. He explains the genetic and evolutionary basis of our behaviour, be it sexual, altruistic, or even reverential.

He responds to his leftist critics frankly. "If history and science have taught us anything, it is that passion and desire are not the same as truth." He responds to his rightist, religious critics just as frankly "Blind faith, no matter how passionately expressed, will not suffice. Science for its part will test relentlessly every assumption about the human condition."

In 1998 Wilson published *Consilience: The Unity of Knowledge,* a treatise on ways the gulf between the humanities and science can be reconciled, partially based on an analysis of how the diverse sciences have found ways to feed each other and find common ground. This complex and fascinating book is refreshingly interdisciplinary in a time of extreme specialization. It also shows him as more than willing to accept the role of culture and environment in what makes us human, despite the popular misrepresentation of him as rigid biological determinist. Relevant to one of my own central themes is his idea that while art is *not* a part of inherited human 'nature', appreciation of art or the drive for aesthetic experience is. He also maintains that there is no inherent reason that these aesthetic or 'spiritual' needs cannot be a valid subject for scientific investigation.*

E. O. Wilson is an articulate proponent—and fine example—of the way science and art can work together toward their common goal of apprehending and appreciating the world. Unfortunately his story is an example of the still present and often nasty resistance to rational reconciliation and consilience.

---

* Obviously, I wouldn't be writing this book if I didn't agree.

THE IDEA OF EVOLUTION; THE EVOLUTION OF IDEAS

"It is nice to believe truth outs in the end. But this is based on two assumptions. One is that with ideas, as with organisms, survival of the fittest is the rule. The second more questionable assumption is that truth is more fit than reassuring fiction. Certainly fiction gets more exercise?"
—Hippokrites

.

"If I were to give an award to the single best idea anyone ever had, I'd give it to Darwin, ahead of Einstein and Newton and everyone else. In a single stroke the idea of evolution by natural selection unifies the realm of life, meaning, and purpose with the realm of space and time, cause and effect, mechanism and physical law. But it is not just a wonderful scientific idea. It is a dangerous idea."
—Daniel Dennett (*Darwin's Dangerous Idea*)

Evolution *is* a dangerous idea. However, all ideas are dangerous. Ideas imply action, be it purely mental or emotional realignments or actual physical action out in the world. Action means change, and if everything is going swimmingly, the chances of change being for the better are slim. If it ain't broken, don't fix it. Messing with the status quo is always dangerous.

.

When I hear my wife say she has an idea, I cringe. I'm not suggesting* she doesn't often have good ideas at a better than 50/50 chance level. It's just that I know that more often than not her idea involves me eventually doing something. And, I admit it, I'm lazy and excessively fond of the status quo (homeostasis is my friend) and really don't want to redecorate the kitchen or completely restructure a book I'm working on according to a new, more sensible outline she has thought up. All ideas are disruptive. Some are so very disruptive that the adjective 'dangerous' is definitely justified.

.

---

* I don't dare!

New ideas are coming at us fast and furious. One of the many things the *Bible* got wrong is that "there is nothing new under the sun".[*] However, it is true that all new things do evolve from old things. They don't spring into spontaneously into existence, with possible exceptions being Athena from the head of Zeus or the universe from a singularity in a big bang. The idea of evolution by natural selection itself evolved from earlier ideas about evolution. The sixth century B.C. natural philosopher Anaximander of Miletus is credited with first proposing the idea of evolution or transmutation of species.[†] Darwin's brilliant, revolutionary new idea that evolved from previous ideas was the *mechanism* of this change: natural selection. But even revolutions are a result of evolution.

.

The speed of evolution for ideas is far greater than for organisms. This is demonstrated by the extent to which Darwin's dangerous idea has evolved into the contemporary science of biology and the contemporary technology of bioengineering in the mere century and a half since the publication of *On the Origin of Species by Means of Natural Selection.*

.

It is the nature—and speed—of this evolution of ideas that is relevant to human creativity.

---

[*] In Ecclesiastes, where some things are *very* right; e.g., "To everything there is a season."

[†] According to the Roman writer Censorious, "Anaximander of Miletus considered that from warmed up water and earth emerged either fish or entirely fishlike animals. Inside these animals, men took form and embryos were held prisoners until puberty; only then, after these animals burst open, could men and women come out, now able to feed themselves."

MEMES, MIMESIS AND MEMORY TRACES

The scientist and philosopher Richard Dawkins (of *The God Delusion* fame and infamy) has developed a theory of the evolution of ideas based on biological evolution. He coined the term 'meme' as the ideational or cultural analog to the gene in his 1976 book *The Selfish Gene*. Memes are intended to represent cultural entities that are passed on from generation to generation according to the same—or very similar—laws of natural selection that apply to the transmission of genes. Although some critics have suggested that a meme is nothing more than another superfluous word for 'idea', it really implies more than is usually meant when one speaks of ideas. It includes fashions, tunes, superstitions, art movements, ways of doing things, belief systems, what sociologists call 'mores' and 'folkways', proverbs and clichés, and—well you name it. Whatever is not covered by genetic transmission, but clearly passes down through generations, has to have some mode of transportation. Dawkins proposes the *meme* as the vehicle. To have such a term seems reasonable to me—and useful for discussing where art and science are evolving.

Before considering the implications of this concept, it is worthwhile to review the three basic principles of natural selection.

- Variability
- Modifiability
- Competition

Variability simply means that there have to be differences between the entities involved in evolving. In biological terms this means that the offspring cannot be identical to the parents. In sexual reproduction the random pairing of chromosomes from both parents guarantees this.

Modifiability in genetics means that the genes can be modified by external events or environmental accidents such as mutations.

Competition is the key component. Given variability and modifiability, the results (offspring) of reproduction will compete for survival and reproduction. This is the much misinterpreted and misapplied 'survival of the fittest' idea.

Completely discredited is Lamarck's idea that modifiability of the phenotype (i.e., the expression of whatever genes an organism has inherited) can be inherited. Giraffes didn't develop long necks from their parents and parents' parents constantly stretching to get the good munchies high up on the trees. I can pump iron everyday, but it isn't going to give my sons a head start on big muscles. But how come it is that dads who are body builders often have sons who are muscular too? Why were many of Johann Sebastian Bach's sons so musical?

Again the answer partially resides in understanding the relationship between nature and nurture. Nature supplies the potential. Nurture supplies the stuff to actualize that potential. We have a handle on how nature works: it is called genetic and evolutionary theory. But how exactly does nurture work so as to pass on specific nurturances from generation to generation to generation. How do cultures, ideas, memes, evolve? And why do they seem to evolve so very much faster?

And, very importantly, what is it that defines 'fittest', and thus survival, in this case? Are the memes that are most likely to survive the 'best' in terms of our humanistic ethical and aesthetic standards? Are the best innovations, the highest moral principles, the best art, the most significant creative ideas those that will move on down through future generations—or is that just wishful thinking. Is it the same wishful thinking that leads some to believe *Homo sapiens* are the best possible outcome of biological evolution?

The good guys don't always win. One example of this in terms of technology, consider the meme of videotape technology. The experts agree that the beta format was superior to the winner technology of the VHS format.* Dawkins, despite his campaign against religion and superstition, would have to admit that even the most patently absurd beliefs have incredible staying power.

So what is happening here?

Variability is guaranteed by the fact that the diversity of memes is huge. I'm annoyingly fond of remarking that ideas are a dime a dozen actually even cheaper, in fact usually free and worth every cent. It's good ones that are scarce.

---

* If you want a chorus of agreement about the inferior technology winning out, talk to anyone who has used any computer operating systems other than the ubiquitous Microsoft OS.

Modifiability is easier with memes than with genes. No mutation-producing radiation is required. We're closer to Lamarckian theory here. If I promote—or force a regime of—pumping-iron at the gym (or going to church) on my kids I can modify their behaviour, and they can—and likely will—pass that behaviour on to their offspring.

Competition? Well there's the rub for those who want to believe the good guys always win. The winners are those who through largely random and unpredictable events win out in a short term competition. Herein lies the explanation for the speed and apparent irrationality of memetic, as compared to genetic, evolution. An analogy is the rolling of a pair of dice. Monsieur Seven will be the long term winner, the fittest, for he has a probability of coming up tops 6 out of 36 times, higher than his closest competitors, Messieurs Six and Eight, which each have probabilities of only 5 out of 36. But if a game ends after only a few rolls of the dice, Monsieur Seven most likely won't be the winner. He needs time to prove his superiority, for the rules of probability are only accurate over extended repetitions. If, as in evolution, especially memetic evolution, there are no rematches, the long-term fittest meme may not be the winner. No wonder the history of civilization is such a drunkard's walk, and unlike biological evolution such a fast-paced and erratic walk.

'Mimesis' is a term in traditional literary and artistic scholarship that refers to art imitating life or nature. A 'memory trace', sometimes called an 'engram', is a term for the elusive neurological location of a memory.*

Maybe Jung's scientifically discredited "collective unconscious" can be partially redeemed by the concept of memes. Art and science are not only ways of imitating, emulating, simulating, and explaining life, they are ways of embedding life in some kind of collective memory trace. As with the neurological memory trace, it clearly isn't neatly stored in one place like a gene is stored in one segment of some chromosome: it is stored in a *network* on connections. Parts of this network are physical, such as libraries and museums and files on

---

* A somewhat specific location for some simple engrams having to do with classical conditioning has been isolated in the cerebellum, but generally speaking the scientific consensus is that most memory is stored in a complex network of connections spanning very distant parts of the brain.

computer hard-drives, but any given meme ultimately exists in the abstract realm of inter-relationships.

As with biological evolution, there is constant branching occurring. Like biological evolution, extinction is possible. However, it is different in important ways, this evolution of ideas, memes, knowledge—or whatever one chooses to call 'it'.

One important difference is that the ideational, memetic or cultural evolution doesn't involve separation after branching. The *tree* is the standard metaphor for biological evolution. The *network* is the metaphor for memetic evolution. (I will hijack and henceforth use Dawkins' term in the following discussion, although he might quibble with my usage of the term meme and may even disagree with my conclusions.)

There are other differences in degree of the three key characteristics of evolution by natural selection: in memetic evolution *variability* and *modifiability* are probably greater, and certainly faster, while the results of *competition* are less drastic for the losers. I propose the following four important characteristics of memetic evolution, relevant to changes in the arts and sciences and which are different from biological evolution and have important future implications.

- *Not about winners and losers*: Memetic natural selection has less to do with a competition in which there is a winner and a loser.
- *Not about branches but rather about networks*: Things do not become separated when a meme branches, but instead remain connected in complex ways.
- *Not slow but very fast*: Memetic evolution proceeds at a rate far, far greater than biological evolution, at least in higher organisms.
- *Not as physical and local*: Physical 'storage facilities' are less important than in the past and are becoming even less important.

Winners and losers. If two species have to compete for a particular ecological niche which has limited resources, there will a clear winner and a clear loser. It will be a duel to the death—perhaps the death of a species: extinction. The recent increase in species extinction and decrease in biological diversity is justifiably worrying. Human beings are largely to blame, unlike other periods before we walked this earth where biological diversity was decreased by natural events such as an asteroid strike.* What is happening now may be a nasty twist on winning the battle but losing the war. We are the short-term winners in competition with other species competing for the same resources, but too many of these short-term victories may result in our ultimate defeat. Take no prisoners is a bad policy from an ecological standpoint.

However in terms of memetic competition, co-existence is more common. The current trend is for greater diversity, not less, and it seems likely to continue—unless, of course, our misbehaviour in the ecological sphere does us in. The situation is different in science than in art, but an increase in diversity still seems to be the trend.

In science, two or more ideas may compete so that in the end there is a winner who effectively wipes out all its competitors and the idea, or meme, becomes extinct. Currently there are numerous competing theories regarding the extinction of the dinosaurs. They include asteroid impact, massive volcanic eruptions, changes in the earth's orbit, a greenhouse effect, mammals eating dinosaur eggs, and a super-nova bathing the earth in deadly radiation. At present the first of this list seems to be the dominant theory, but it will co-exist and compete with the others until some overwhelming evidence confirms it or discredits all its competitors. I wrote earlier about competing theories regarding the origin of life on earth. These are examples of the two effects of learning more through the scientific method. One effect is to increase the number of reasonable hypotheses (and so increase memetic diversity), and the other is for a further increase in knowledge to eliminate hypotheses previously considered reasonable (thus decreasing memetic diversity). The former effect has consistently been greater than the latter. For every hypothesis discredited by new data, several new hypotheses are needed to try to explain the increased complexity those data introduce. Things just keep getting more and more complicated in the domain of science at the micro level, even as things gets simpler and more unified at the macro level.

In art, the competition is less bloody. Competing memes just go on competing. One may achieve dominance, but the vanquished aren't

---

* Humans weren't around and so clearly can't be blamed for the rapid and catastrophic mass extinction of species 65.5 million years ago called the "K-T Extinction Event".

put to eternal rest—as happens when a scientific hypothesis is proven wrong. Art movements such as Impressionism may rise to prominence and then lose *cachet* and come to be considered *passé*, but they aren't then put to death. (A Monet painting of water lilies recently sold for 36.6 million dollars at a Sotheby's auction. Anyone who claims Impressionism is dead is brain dead.) The evolution of art is the topic of the next section, so suffice it to say here that art memes proliferate and rarely are killed off by competitors.

Networks Instead of Branches. Biological evolution is more fragile than memetic evolution because of its hierarchical structure. When whatever it was that cut off the evolutionary branch that had the great dinosaurs balancing at its tip, that was *it*—the end of the road for the big guys. They had no sturdy lifeline connected to another branch on which to swing their huge hulks over to a new perch on which to go on with their evolution.

Because of this, memetic extinction is far less likely. As I've repeatedly remarked, creativity is about seeing and exploring relationships. These relationships are not hierarchical, but rather a complex network. Networks are the most resilient and indestructible things imaginable. The Internet is a perfect example (and one to be discussed in detail in the next section), for there is no one branch that can be cut off and so cause everything beyond it to fall to its death. There are lifelines connecting everything to everything. Like the human brain, where each one of our ten* billion neurons has two thousand connections to other neurons, cutting one connection is trivial and cutting enough to seriously disrupt communication or wipe out memory or meme requires a massive attack, such as (with the brain) a massive stroke. Even in the case of a stroke, the brain, like all networks, shows incredible plasticity. This is because other parts of the network peripherally connected to the destroyed area take up the chores of the dead cells.

This interconnectivity means that both scientific and artistic memes remain fertile and capable of sprouting new branches long after they have been lying dormant and in the shade of more thriving and growing branches. An example in science is those "monsters", fractals (known of but ignored for centuries) suddenly experiencing a growth spurt when stimulated by the meme of fascination with

---

* This a most modest estimate. 100 billion has been suggested.

computer technology's potential for iteration and visualization. An example in art is the surrealist style of the paintings of the 15th Century artist, Hieronymus Bosch, becoming the major art movement of the 20th Century.

.

Faster and Faster. In general, biological evolution takes time—lots and lots of time. It is fair to say that close to five million years is a long wait for complex *us* to make our debut on the world stage. Fortunately Mother Nature is very patient, literally having all the time in the world. This is why the study of the process is mostly done *ex post facto* through the fossil record and not experimentally. Nevertheless scientists decades ago have replicated and observed biological evolution, contrary to the claims of creationists.* They have been able to do so with fruit flies, which are the ideal subjects for genetic experiments because scientists can observe a hundred generations in a year's time.

.

Memetic evolution proceeds at a rate far, far greater than biological evolution. The length of a generation of some memes is far less than even that of a fruit fly gene, and certainly less than a human generation. Just consider fashion in clothing as a meme. Who hasn't observed many mutations of it in their lifetime? It is the same in science and art. What neuroscientist can't chuckle when remembering how, for example, the hypothalamus as fashionable research topic was so quickly replaced by the amygdala and now the new 'pleasure centre', the nucleus accumbens? What artist hasn't seen numerous art movements replaced by the latest 'in' approach to creating art?

.

The working time-scale for understanding evolution in science and art is more human and historical than for biological evolution, which is geological or even cosmological. And it seems to be increasing exponentially.

.

Non-locality and Physicality. Physical storage facilities are less important than in the past. This is a result of the network nature of the storage of memes, and again implies the important principle that memes are more resistant to extinction. The gene for a particular characteristic can disappear entirely as one branch of the evolutionary tree is chopped off. Most memes have so proliferated that they are stored in so many places that they are unlikely to ever be lost to

---

* T. Dobzhansky, & O. Pavlovsky, "An experimentally created incipient species of Drosophilia", *Nature* 23, p. 289-292 (1971).

posterity. The destruction of the Library of Alexandria in historical times was the memetic equivalent of the K-T extinction event 65.5 million years ago that wiped out innumerable genes. Such an extinction event for memes is highly improbable today, short of a world-wide catastrophic event that effectively wipes out the human race. The Internet is the new Library of Alexandria, and it is not stored in hardcopy in one place and thus vulnerable to extinction. Shakespeare's works, for example, are so replicated that the whole human race would have to be extinguished for them to be lost.*

However this is a central theme of the next major section, so I won't elaborate here. Suffice it to say that reproductive technology† has so advanced that even the meme least successful in the ideational competition is still out there, safe in numerous, albeit obscure nooks and crannies, just waiting for the opportunity to move out into the spotlight.

Does all this imply that memetic evolution in art and science is quick, quixotic and chaotic, as opposed to the relatively more leisurely and much neater hierarchical, biological evolution we do have some understanding of? And again the teleological monster raises its head to ask if there is some goal toward which either is moving. Is or is not progress a chimera?

---

* The dystopia described in Ray Bradbury's *Fahrenheit 451*, where books are burned and their contents only kept alive by people memorizing them could never happen now. It's now impossible, for not only all the books, but all digital storage media would have to be destroyed.

† I have to smile whenever using that phrase. Of course I'm not talking about biological, genetic reproduction, but rather memetic reproduction—not about sex but rather about printing, digital recording, etc.

ARE ART AND SCIENCE EVOLVING, OR ARE THEY JUST REVOLTING?

Too many people do find much of art and science 'revolting', but presuming they aren't reading this, I'm going to concentrate on the question of whether art and science are evolving.

.

Among those willing to concede the obvious fact of evolution, some do so only because they choose to misinterpret it as a process leading up to *Homo sapiens*. Like the concept of progress, evolution is sometimes misconstrued as implying an inevitable path leading to the ultimate destination of perfection: *us*. (Humans certainly haven't lost their sense of self-importance or self-esteem, despite all the very good reasons to do so.) We have no right to claim we represent the pinnacle of 'evolutionary progress'. Sure, we're not doing all that badly for the moment, but there are many species that have been doing just jim-dandy for millions of years before us and are resilient enough to keep on keeping on even if we bathe the planet in radiation from a nuclear war. The 'modern cockroach' has been around for 150 *million* years. Talk about seniority and staying power: 'modern man' is a newbie here with survival for, at best, a mere 130 *thousand* years to his credit. And since cockroaches can survive radiation levels several magnitudes greater than *Homo sapiens* and generally thrive in environments that would kill you or me, a smart gambler would put his money on that green-blooded little critter over that red-blooded creature proud of inventing Raid. Sad to say, a far better case could be made for the cockroach as being the epitome of evolutionary progress than for us.

.

However, there *is* progress and there *is* evolution, even though there *is no* evidence or reason to believe that either will end in some final perfection and completion—in any sphere. What there *is* evidence for is that it will *all* eventually end in the great heat death and boring cold entropic nothingness, but that is too depressing to contemplate— except briefly as a bracing tonic to wake us up to reality when we're feeling too uppity. In the fortunately extensive meantime, it is interesting to speculate whether or not art and science do evolve in a way that could be called progress.

.

To deal with science perfunctorily, let me say outright that progress through evolution is an obvious indisputable fact. Our scientific understanding of the workings of the universe is unquestionably better and deeper than it was in the past. Certainly we may now wrongly reject something that once was considered—and was in fact—true. But in time that will be corrected, for science is self-correcting. Of course it is unjustifiable to think we'll eventually solve and explain all the mysteries of the universe, but the idea of progress doesn't imply that. And who can seriously deny that we have made progress in scientific understanding? Who can deny that the memes of science have evolved in a way that has made our vision of the natural world clearer and deeper?

.

So what about art? Does it progress? Does it evolve? Here it should be emphasized that these are two different questions. Progress and evolution are not synonymous.

.

Progress implies things getting 'better' or more complete. As already emphasized, this doesn't mean things will ever get from better to some ultimate *best*. But it is fair to say that learning the first one thousand decimals of pi is progress over only knowing the first ten—even if we will never know all of the numbers after the decimal to the infinity to which we suspect pi extends.

.

Evolution does not imply things getting 'better'. Evolution is about change and adaptation to changing circumstances. So it may imply things often getting more complicated and more complex, but that is an entirely different thing. More complicated and complex things are more easily broken.

.

It is certainly true that judgments are routinely made where someone says one work of art is better than another. However, such judgments are not based on the work's place in that art's evolutionary history. Is Mozart better than Bach, and is Beethoven better than Mozart, because Western Music evolved from the Baroque style to the Classical Style to the Romantic Style? I think both Mozart and even arrogant Beethoven would admit their predecessor Bach was their superior. And certainly very few composers since the evolution from the Baroque style would have the *hubris* to claim they are better than Bach.* Are monophonic Gregorian Chants inferior to homophonic hip-hop music played on AM radio because the latter evolved from the former? Theatre has evolved greatly since Shakespeare's time, but who would say it has *progressed*—and so go on to claim that The Bard

---

* While what I'm saying is indisputedly true, I have to confess to a bias. I've been known to (after a few too many drinks), to pontificate on how music has "been all downhill after Bach".

was just a sort of *Homo habilis* playwright, a mere precursor (with his knuckles dragging on the ground) to the more evolved *Homo sapiens* playwrights now writing scripts for Broadway productions?

.

Art changes. Art evolves. Art adapts to changing circumstances and new environments. Art incorporates new tools (often imported from science), but often the actual works made with 'primitive' tools are superior to the new works made with the new-fangled tools. New art is different from older art, but different isn't necessarily better. We wouldn't have Beethoven's "Hammerklavier Piano Sonata" if it weren't for huge improvements in the fragile pianos of Mozart's time made by the firm of Broadwood (and gifted to 'Louie' for experimentation)—and if it weren't also for Ludwig's personal commitment to the meme of 'art as personal expression'. So bless change, bless science, bless technology, bless memetic evolution. But that doesn't make this brilliant "Hammerklavier Sonata" better than Bach's keyboard sonatas, which had to be played on a harpsichord that couldn't express the dynamic range* required for Beethoven's sonata.

.

Evolution in science allows us to *know* more, and so one can justifiably call that 'progress'. Evolution in art allows us to *do* more, but progress is an inappropriate, or at least misleading, term: 'change' is more accurate.

---

* The harpsichord, unlike the piano, creates every note at the same volume. Hammering a key won't increase its loudness.

ANY CHANCE OF EVOLUTION IN RELIGION?

I've already discussed in various contexts the role religion plays in creativity, and the reader probably has the impression that I think that that role is that of villain. This isn't entirely accurate. While it is my opinion that religious beliefs *in general* (and especially when extreme) have done far more harm than good, this isn't to say they haven't sometimes, even often, been a means to a worthwhile end. And when as means they are innocuous, and the end is a creative one, they have to be appreciated.

.

The evidence is overwhelming that Johann Sebastian Bach was deeply religious. Even the little 'throw-away' pieces he composed for his children he always dedicated to God. Bach reputedly called music "the apparatus of worship".[*] If his religious belief was integral to his creativity, as it certainly seems to have been, then it would be insane to wish he had had no Faith. The reason secular humanists so often rant about the evils of religion is because it is entirely reasonable to wish that potential suicide bombers had lost their Faith before they could strive for martyrdom and some heavenly reward by killing themselves and lots of other people. But who in their right mind would think Bach losing his Faith would be a good thing if it meant Bach *St. Mathew's Passion* would never have been written?

.

It is fair to say that historically far, far more often than not, religion has played some role in the creation and appreciation of art. The great medieval cathedrals weren't constructed by atheists for the purpose of inducing aesthetic experiences. They were built for "the greater glory of God". And the appreciation they received at the time was religious, not purely aesthetic.

.

A few years ago I conducted a research study with my then student Shelley Taylor. Her hypothesis was that because the descriptions of profound religious/mystical experiences so closely matched those of profound aesthetic experiences, the alleged difference may only be the result of a personal experiential bias, of a personal cognitive schema. In other words what was happening was a simple case of labelling differently the same phenomenon, depending on whether one was religiously or artistically inclined.

.

---

[*] This according to the entry on Bach in the *Grove Encyclopedia*.

We had over five hundred respondents to a questionnaire we constructed to address this question.* Among those respondents who met the criteria questions common to descriptions of both religious/mystic and profound aesthetic experiences we found a significant correlation between religious or artistic orientation and the claim to have had a religious/mystic or profound aesthetic experience.

.

As I remark in the conclusion to a paper reporting our findings†, "The feudal peasant travelling from his humble abode to attend Mass at the majestic Köln Cathedral surely was moved profoundly by the grand architecture, the gorgeous stained glass windows, and the magnificent music played on a giant pipe organ, and he surely felt this to be a religious experience. However, the contemporary unreligious, but aesthetically engaged, tourist is likely to have a very similar experience when first visiting this grand cathedral, but instead interpret it as a profound aesthetic, rather than religious, experience."

.

Of course our study does not prove that religious/mystical experiences are nothing more than mislabelled profound aesthetic experiences, anymore than it proves the reverse. It does, however, suggest that art and religion can find common ground on which to stand and view with awe and reverence the world we live in and our creations.

.

.

In the earlier chapter on what I've called the "Psychic Circus", I expressed some optimism regarding the reintegration of science into art resulting in artists being less susceptible to humbug. I also expressed some pessimism, or at least concern, regarding the general social trend toward increased willingness to embrace humbug and irrational religious belief. Is there a contradiction here?

.

If I am correct that there is a trend for art and science to come together again, I would speculate that there is another similar trend: art and religion resolving some of their differences as well. Now it is

---

* Approximately half of these were university students, and half were respondents to the questionnaire posted on a website devoted to the study.

† Stange, Ken, Taylor, Shelley. (2008). "Relationship Of Personal Cognitive Schemas To The Labeling Of A Profound Emotional Experience As Religious-Mystical Or Aesthetic". *Empirical Studies of the Arts*, **26(1)**, 35-47.

well known that artists interested in shocking their audiences find the religious community an easy target. There is a 1926 painting by the surrealist Max Ernest showing the Madonna Mary spanking the Baby Jesus and His halo flying off. More recently (in 1989) Andres Serrano exhibited a photograph, entitled "Piss Christ", of a small plastic image of Christ on the cross submerged in a glass case allegedly filled with the artist's urine. Needless to say, both works caused considerable controversy.* This sort of believer-baiting has aggravated the extreme animosity toward art that one associates with those who are fervently religious. So how can one possibly conceive of any kind of reconciliation?

.

It is unfortunate that more and more it seems both science *and* art are seen as the enemy by the devout, for the devout are legion and their power to suppress creative expression frightening. Scientists have always, albeit almost always unintentionally, upset the religious establishment and the fanatically devout. But for much of history artists were actually supported by the Church. The Church was *the* patron of the arts through most of the history of Western European art. Writers occasionally got into trouble, admittedly, but overall the relationship between artists and religious leaders was symbiotic, mutually beneficial. Pope Julius II financed Michelangelo's work on the vault of the Sistine Chapel, and the awe his ceiling produced in the worshippers served to reinforce belief and the power of the Papacy. Whether Michelangelo was a true believer is really irrelevant.

.

But the church is no longer a major patron of the arts, and most art is no longer, even putatively, created for the "greater glory of God" (and to increase religiosity in the laity.) It should also be noted that science and scholarship too were once (for at very least a thousand years!) also under the patronage of The Church in Europe. This responsibility for financial support of both art and science has passed into the hands of government and private enterprise. Art is created 'for art's sake'—and that the artist's work not offend is only a practical concern regarding granting agencies and the marketplace.†

---

* Incidentally, they are an example of how shock value has come to take precedence over aesthetic value. Ernst's painting is witty and quite aesthetically interesting. I'd love to own it. I can not say the same for "Piss Christ".

† This is not a trivial concern. Offending government, or private patrons, or a public obsessed with being politically correct may even be more likely than it was during the days of Church patronage.

Having said all this, which may seem an argument against my thesis of reconciliation, I still maintain that there is some light on the horizon. If in fact the secret agents of science are infiltrating the artistic community, they also are infiltrating the religious community. The profound personal experience that great art produces has always been desirable, sought out, and rewarding. Religious institutions would have collapsed a long time ago without such experiences to prop them up. Even the most fundamentalist and 'puritan' Christian sects that consider most music the work of the devil still have choirs.* The joyous frenzy of gospel singing at a Baptist prayer meeting is not different in kind from a rave dance concert. If science and reason can temper the intolerance too typical of religion, then at least some artists may be welcomed back into the fold. There is a genre of rock music called "Christian Rock", and even radio stations devoted to it. I know, I know, the stuff is pretty lame, but that fundamentalist Christians can open up to any kind of Rock music is a positive sign.

Call it religious/mystical experience or call it aesthetic experience, call it whatever you want. It is an experience that will be sought out and one that will get converts—to religion or to art. Secular humanists may wish to see the demise of all religion, but that just isn't going to happen in the foreseeable future. But if the respect for reason and empiricism that distinguishes science can temper the extremism and intolerance that too often is associated with religious belief, there is hope that there will be more Bachs inspired to write beautiful music and fewer Jihadists inspired to murderous atrocities.

---

* Spirituals are one of the roots of their demonized Rock 'n Roll.

CASE STUDIES: RICHARD DAWKINS, JOHN COLTRANE

Clinton Richard Dawkins (born March 26, 1941 in Nairobi, Kenya) needs no introduction, for currently he is a very high profile combatant in the so-called 'science wars', and I have already mentioned him several times in this book. John Coltrane (born September 23, 1926 in Hamlet, North Carolina) also should need no introduction, for since his death in 1967 he has attained legendary stature as one of the great masters and innovators of the unique music genre called jazz; with even a commemorative U.S. postage stamp in his honour being printed in 1995.

.

Coltrane was as religious as Dawkins is irreligious, but Coltrane's influence could be cited by Dawkins as an example of a memetic evolution—a positive memetic evolution. The optimist in me likes to think that in religion, as in jazz, the evolution seems to be moving toward greater freedom and diversity. If so, John Coltrane is a perfect example of this.

.

He grew up in a segregated southern American town long before the civil rights movement. Both his grandfathers were ministers in the African Methodist Episcopal Zion Church. His mother was the church's pianist; his father played violin. Before financial disaster struck with the deaths of several of the extended family's bread-winners, his was an exceptionally middle-class childhood. This meant, for an African-American family at the time, a strong emphasis on religion; and, because music was so central to religious ceremony, a musical education that included the Western European choral canon. The young 'Trane', as he was to be called most of his life, played alto horn, clarinet, and alto saxophone in various school, community, and church bands—and sang in the church choir. This early close association between the emotionally charged domains of music and religion was to surface later in his life when he came to believe that music was more than just music: music was a mystical language that expressed the deep, spiritual core of human existence and the universe.

Coltrane had discovered jazz near the end of his high school years, and since at the time this was *the* music of the black community and was also so fascinatingly complex and sophisticated, naturally he fell in love with this new genre. Coltrane was drafted into the Navy in

1943, where he played in a band that played bebop, the current, trendy movement in jazz.

.

Upon being discharged from the navy in 1946, he worked various jobs and played various gigs, until in 1949 he was welcomed into the famous Dizzy Gillespie Big Band. The jazz life scene at the time included easy access to drugs, and he became addicted to heroin. When Gillespie's group disintegrated, Coltrane played with various groups and free-lanced until 1955 when another jazz great and heroin addict by the name of Miles Davis recruited Coltrane for his quintet. It's a cliché, but the rest *is* history—jazz history.

.

Miles and Trane were the Deadly Duo superheroes of jazz fans at the time—and ever since. In 1957 Coltrane kicked his habit, largely through a sort of religious conversion to Islam initiated by his first wife Naima, who held Muslim beliefs.

.

For whatever reason, belief in God (in whatever guise) seems to be commonly associated with overcoming addiction. (Many unreligious people wanting to beat an addiction are repelled by the methods of the Alcoholics Anonymous organization because of its Twelve Step program which is entirely based on belief and subservience to "a Power greater than ourselves.") No matter what one thinks of religion, in Coltrane's case the crutch of renewed religious belief helped him get back on his feet.

.

From then on until his death ten years later at the age of forty*, his belief in spirituality was again inextricably bound with his music, as it had been in his youth, and always had been in the black religious community. The parallels with Bach (and many other baroque composers) in this regard are obvious. Music was not music for music's sake. Music was spiritual with a higher purpose. It was not mere entertainment—even serious and important and enlightening entertainment. It was a form of communication with God, the

---

* A personal anecdote. In March 1966 (about one year before his death) Coltrane performed at The Plugged Nickel club on Chicago's Rush Street. I was a university student at Loyola at the time and under drinking age. However, like many of my fellow students, I'd acquired fake ID indicating I was the drinking age of 21. Unfortunately, such fake IDs were so common that savvy club owners limited admission to those with IDs indicating the kid showing the ID was at least 23. So I missed a chance to hear the master. (I was older and not in need of fake ID when I did have the opportunity to hear that other great jazz artist Miles Davis live in a Toronto club in 1969.)

Cosmos, or the noumena beneath phenomena. In the liner notes for his 1965 album *Meditations* he wrote that the purpose of his music was to inspire people "to realize more and more of their capacities for living meaningful lives. Because there certainly is meaning to life."

.

This meaning he sought out and seemed to find in almost every imaginable religion. He was interested in and researched Sufism, Hinduism, and Buddhism, using references to the sacred texts of these religions in various musical works. He was eclectic and voracious in his search for this meaning and also turned to the ancient Greek philosophers and research into math and science, as well as the less credible areas of astrology and mysticism. He may not have been coolly, rationally critical in his appraisal of the answers offered (he may even be said to be naïve*), but he was certainly tolerant and open to any and all explanations of our ultimate *raison d'être*.

.

It is hard to fault this brand of religiosity and spirituality which embraces so much and seems untainted with any of the nastiness of dogma. This is especially true if it is to be credited with both his recovery from addiction and the creation of the innovative and beautiful music he created in the ten short years from his 'conversion' up until his untimely death. I remember the first time I heard his classic "A Love Supreme" and my profound reaction, a reaction I still have now and which is strikingly similar to my reaction to Bach's choral music. It is a deeply moving hymn, unattached to any specific religion or dogma, that should make even the most anti-religion atheists pause and wonder if there may be something lost if we should all become rational secular humanists.

.

This is, of course, a question to ask Richard Dawkins, whose book *The God Delusion* is a full frontal attack on religious belief in all its

---

* A thesis posted on the Web by Scott Anderson, "John Coltrane, Avant Garde Jazz, and the Evolution of "My Favorite Things" well describes Coltrane's beliefs regarding the power of music. "Coltrane's study of Indian music led him to believe that certain sounds and scales could 'produce specific emotional meanings' (impressions) According to Coltrane, the goal of a musician was to understand these forces, control them, and elicit a response from the audience. Like Pythagoras and his followers who believed music could cure illness, Coltrane said: 'I would like to bring to people something like happiness. I would like to discover a method so that if I want it to rain, it will start right away to rain. If one of my friends is ill, I'd like to play a certain song and he will be cured; when he'd be broke, I'd bring out a different song and immediately he'd receive all the money he needed.'"

manifestations. Thomas Huxley's vigorous defense of Darwin earned him the appellation "Darwin's bulldog". Dawkins advocacy of evolutionary theory has earned him the epithet of "Darwin's Rottweiler". And religious fanatics bypass reference to any specific breed and just consider Dawkins a "mad dog" trying to rip to shreds their most treasured beliefs. The facts are that bulldogs are quite a placid breed, that Rottweiler's reputation as vicious is unfounded, and that Dawkins certainly is not mad, and his major similarity to dogs is his gentle nature and affection for humanity.*

Dawkins was eight when his parents moved from Africa to England, where eventually he studied zoology at Oxford, where he studied with the Nobel-prize winning ethologist, Nikolaas Tinbergen. I've been unable to find any reference to him being particularly disputatious or abrasive as a student. He received his MA and DPhil degrees in 1966 and the following year was hired to teach at the University of California at Berkeley.

It was only with the publication of *The Selfish Gene* in 1976 that he became famous outside the scientific community. As previously mentioned, it was in this book he coined the word "meme". This was followed by *The Extended Phenotype* (1982) and *The Blind Watchmaker* (1986). By this point in his career he had been put on the defensive because his clearly defended ideas offended believers in some Grand Design with God as architect drawing the blueprints—and also because his scientific arguments for the premier role in evolution being given to the gene were disputed by other evolutionary theorists, including the eminent Stephen Jay Gould. His opponents in the religious camp were nasty and irrational, but those scientists who disagreed with him were in most cases temperate and reasoned in their criticisms. Opponents in the religious camp were nasty and irrational, but those scientists who disagreed with him were in most cases temperate and reasoned in their criticisms.†

---

* Although he has been known to occasionally (albeit rarely) snap at those who attack him.

† Stephen Jay Gould is the most notable of his opponents in the scientific squabble, and Dawkins (like his friend Daniel Dennett) always maintained a cordial relationship with Gould, and while disagreeing with him, consistently publicly praised his scientific accomplishments. This is an example of scientific debate at its most civilized. Unfortunately, it isn't always so amiable and courteous.

I've already been guilty of cheap psychologizing about Dawkins when I discussed the teleological urge, but it doesn't seem a stretch to suggest that it was the attacks against his really very moderate suggestions in his early books that led to his increasingly hostile analysis of those who would consider their superstitious beliefs as having equal status with reason and science. In reading many reviews of his *The God Delusion*, I've found the majority of reviewers label Dawkins as a proselytizer for the "religion" of science. This is even true of those who seem to basically agree with him. Of course this is nonsense. Science is not a religion, for it differs from religion in the most fundamental way possible: it is based on scepticism, not dogma, and is a self-correcting epistemological method. If one removes God from the equation, it *is* fair to compare Communism, Nationalism, Fascism, and Nazism to religion, for they are, like traditional religions, based on dogmas (not empiricism and reason) that cannot be questioned—but may be 'interpreted' to suit the needs of the higher-ups in the hierarchy. It isn't unreasonable to consider the writings of Marx or Lenin or sayings of Chairman Mao as the holy scriptures of various sects of communism. There is no question that such godless 'religions' do as much harm as do the god-based ones. But the only thing science has in common with them is the rejection of some Supreme Being. Everything else they share with traditional religion.

.

Dawkins' more snarling and rabid attackers have tried to drive him into a corner. It is very much to his credit that he has not responded in kind. Justifiably confident in the—what should be—common sense of his critical observations on religion, he somehow has managed to remain calmly rational—which undoubtedly annoys the hell out of his enemies. His creative disagreeableness seems to remain untainted by any personal animosity toward those who attack him. This is rare in artists, and while less unusual in scientists isn't exactly common when the stakes are high or those attacking are just plain nasty and ignorant. Suffering fools gladly is not common among the creative. Despite his rottweiler reputation, Dawkins is more tolerant of fools (although not foolish ideas) than anyone has a right to expect.

.

However, he seems to be a victim of a meme of his own creation: the idea that religion is a pure evil which has no redeeming value *whatsoever*. I certainly have never found in my reading of him a single kind word about religion. He even dismisses the power of religion to offer consolation—no matter how specious such consolation may actually be. In an interview with Sheena McDonald*, in response to her remarking that people less fortunate than him find comfort in religion, Dawkins replied "There are all sorts of things that would be comforting. I expect an injection of morphine would be comforting – it might be more comforting, for all I know. But to say that something is comforting is not to say that it's true." And what he says is true—but beside the point. There is nothing evil in a comforting untruth. The mother by the bedside of her terminally-ill child telling him he is going to be all right, that God is watching over him, is not doing evil.†

.

Dawkins is a deeply sensitive man as his occasional non-polemical remarks indicate. In that same interview with McDonald he says "The world and the universe is an extremely beautiful place, and the more we understand about it the more beautiful does it appear. It is an immensely exciting experience to be born in the world, born in the universe, and look around you and realise that before you die you have the opportunity of understanding an immense amount about that world and about that universe and about life and about why we're here." In 1991 he was the guest lecturer for the Faraday Lecture Series for young people.‡ His theme was the wonders of science and the world we live in, and at one point he speaks most eloquently about our "privilege" to be here and how great a squandering of this privilege it would be to not try to understand and apprehend the universe we have found ourselves in—if only for one brief shining moment in the immense expanse of time and space. He was close friends with the writer Douglas Adams and shows great respect and appreciation for the arts. Dawkins himself is not only a first-rate creative scientist; he is also a first-rate creative writer. When he writes about the natural world, his prose is imbued with a sense of awe and wonder that— if one really wanted to annoy him— one could call spiritual or mystical.

.

---

* Aired in Great Britain August 15th, 1994 on Channel 4 as part of a series called *The Vision Thing*.

† Obviously this a very different kind of religious lie than telling your son everything will be all right and God will take care of him if he straps explosives to his chest and blows himself up in a crowd of 'infidels'.

‡ The Royal Institution Christmas Lectures for Children were founded by Michael Faraday in 1825. Dawkins lectures are entitled "Growing Up In The Universe" and are available on DVD.

If the cases of John Coltrane and Richard Dawkins can be taken as indicative of the direction the evolution of the memes of religious belief will be taking, there is reason for optimism. If religion memetically evolves in the direction that Coltrane epitomizes, it will be less of the evil influence on civilization than it has too often been, and more of the inspiration that it has often been. And if the unsentimental, clear-thinking of people like Dawkins become even half as common as the vicious and dangerous religious fanaticism that is ravaging our world, then maybe it isn't entirely unreasonable to equate evolution with progress—in some cases.

MATH, THE OLD PATH, NEWLY EXPLORED

"Mathematics, rightly viewed, possesses not only truth, but supreme beauty—a beauty cold and austere, like that of sculpture, without appeal to any part of our weaker nature, without the gorgeous trappings of painting or music, yet sublimely pure, and capable of a stern perfection such as only the greatest art can show."
—Bertrand Russell (*The Study of Mathematics*)

"One cannot escape the feeling that these mathematical formulas have an independent existence and an intelligence of their own, that they are wiser than we are, wiser even than their discoverers, that we get more out of them than was originally put into them."
—Henrich Hertz (Quoted by Eric T. Bell in *Men of Mathematics*)

It has been said many times that mathematics is the language of science. However it would be more accurate to say it is *one* of the important languages scientists use. It may be the primary tongue spoken by physicists—who often have to know several obscure 'dialects'. To continue the use of 'dialect' as a metaphor, consider that while both Cantonese and Mandarin are considered dialects of the Chinese language, and have a common literature and history, they are mutually unintelligible to those who speak only one or the other. As the number of dialects of mathematics increased and became more distinct from each other, even mathematicians themselves confess that they often can't understand what those working in a different field are 'saying'.

Furthermore, in many other areas of science, fluency in the language of math is not required. In fact one can get by with the equivalence of phrase-book knowledge (of a particular dialect) in many scientific fields; some random examples being archaeology, geology, botany, even chemistry.*

---

* Oddly, social scientists need more fluency (at least in one dialect) than do chemists. The dialect I'm referring to is statistics, for human interactions, unlike chemical ones, are not consistent. Two chemicals under identical conditions either react or they don't. This isn't true in psychology where identical conditions are impossible and organisms, unlike chemicals, are not identical. So one needs statistics to estimate the probability of any differences being real or simple chance variation.

What is *not* often said, but is also true in a similar way, is that mathematics is one of the languages *artists* have always used. No classical or renaissance artist was ignorant of mathematical relationships. The obsession with the "golden ratio" (or "divine ratio") that dates back to Ancient Greece is but one example of how mathematics has permeated painting and architecture. And it isn't just in the visual arts that this is the case. The whole Western Musical tradition, back to the octave, is based on the mathematical relationships of notes of different frequencies that were first discovered by Pythagoras while allegedly playing around plucking strings of different lengths. The Pythagoreans were all musicians as well as mathematicians and saw this as entirely natural. And anyone who has studied music theory will confirm that musical notation bears more than a slight resemblance to mathematical notation.* And even the literary arts have a mathematical component. It all began with poetry, in which counting of syllables, beats, phoneme lengths, and maintaining specific ratios of these 'metrics', were the central structural components.

In both art and science, mathematics is a tool. In mathematics, I am told repeatedly by my daughter, it is an end unto itself, not just a means to some other end. Its utility doesn't particularly interest her. She does 'pure' mathematics. There is a parallel here to art. Art for Art's sake. Math for Math's sake. And just as art obviously has many applications and functions, so does math. "Useful? That's nice, but so what?" says the artist or mathematician (or scientist) who simply does what she does for the sheer pleasure of it.

However, for non-mathematicians, be they artists or scientists, interest in "Math for Math's sake" is unusual. Math is merely a tool to them. Similarly, most artists just *use* science (or technology), and scientists *use* art (or aesthetic principles). Only those few deeply committed to both don't think in purely utilitarian terms when drawing on resources outside their creative domain.

I hope there is no need to say that there is nothing wrong with this. It is equivalent to two people helping each other when each has different skills and goals. Jack, the auto mechanic, may derive intrinsic pleasure from diagnosing and fixing engine problems. Jill,

---

* You better know your fractions, kids, if you are going to understand minor thirds and major fourths, your triads and duple meters, etcetera.

the business consultant may derive intrinsic pleasure from diagnosing and fixing financial problems. Jack fixes Jill's car and gives her some tips on proper car maintenance. Jill fixes Jack's business practices and gives him some tips on more effective and efficient ways to run his business. They learn from each other, but in each case they are using the other's skills as a means to an end—not as something intrinsically rewarding.

.
.

This book is about art and science. I'm not sure where to place mathematics. Is it an art or a science? I guess I would say it is both. But what matters in the current context of future developments is that it is of utility to both. Its ongoing utility in science is not at question. Its utility in art is again becoming appreciated, as a spin-off from artists realizing science is useful in their endeavours. This looks like a long-term trend.

SCRATCHING THE SURFACE TO LIGHT THE MATCH

My wife has remarked how when you scratch the surface of almost anything you find mathematics. I'll expand her metaphor and say that this scratching on the rough surface of what we naively consider reality also lights a match that illumines, however briefly, the exposed surface of a higher reality. Whether it be artist or scientist that is striking the match, the illumination has often been brilliant—and sometimes momentarily blinding.

.

It is really no surprise that the Pythagoreans created what amounted to a mystical cult out of mathematics. The inexplicable abstract relationships they discovered are more awe inspiring than even the most wondrous events in the perceived world on the surface.

.

One of my deepest regrets is my failure of will and self-discipline which prevented me from reaching even the lower plateaus on the Olympian mountain of mathematical understanding. What little understanding I have eventually attained only increases my regret, for I can now catch brief glimpses of the brilliantly illumined higher regions.

.

One has to admire those with the endurance to climb to those peaks, just as one envies the view they have earned. One also has to feel pity for those who have never even imagined that such vistas existed.

.

Math, maybe even more than science, has a bad rep in the minds of the artistic community and the general public. The difference is that science is often perceived as downright evil and difficult, while math is perceived as merely insignificant *and* difficult.

.

Presumably no reader who has made it this far has such a naïve view of math, even if just as unsophisticated as I am in the field. Math is not about numbers; it is about relationships. Some areas of math really have almost nothing to do with numbers at all; this could be said of topology, one of the largest sub-disciplines in mathematics. .

.

An important philosophical (epistemological) question is what math is *really* about. Saying it is about relationships is only a superficial answer. Are these relationships representative of anything other than themselves? Like, duh, for example, the real world? Is pure math really pure, untainted by any relationship with reality, other than the coincidental?

Putting aside such philosophical questions for now, I'd like to digress to the nature of the mathematician's creativity. It has some relevance to the role mathematicians are playing—and will continue to play—in the creative endeavours of both scientists and artists. Mathematicians are an odd lot, to say the least. Eric Bell (somewhere in his book *The Development of Mathematics*) remarks something to the effect that mathematicians evolving within a race akin to apes is even more bizarre than chickadees evolving from winged dinosaurs.

How *do* mathematicians' minds work?

Assume two trains on the same track are 120 kilometres apart, heading towards each other at 15 kilometres per hour. Suppose a fly, flying at 60 kilometres per hour leaves the first train, flies to the other, turns around and flies back and forth until the two trains collide. How far will that fly travel before it is squashed between the crashing trains?

Think about this problem for a while before reading on. There is a simple logical solution.

The first train is traveling at 15 kph and the second train is also going 15 kph, so they are approaching each other at 15 kph + 15 kph: i.e., 30 kilometres per hour. We know the rate at which the trains are approaching each other and their distance apart (120 kilometres), so it is easy to find the time it will take until they crash: i.e., 120/30 hours, or 4 hours. Obviously the fly is spending the same amount of time traveling as the trains, and if he is flying 60 kph, he will cover 240 kilometres in those 4 hours.

This problem was presented to the brilliant mathematician John von Neumann. He responded with the correct answer in a matter of seconds. The mathematician who proposed the problem grinned, impressed with von Neumann's quick insight into the heart of the problem. "Interesting," he said. "Most people try to sum the infinite series."

"What do you mean?" von Neumann replied. "That *is* how I did it."

This is the man who allegedly said "If people do not believe that mathematics is simple, it is only because they do not realize how complicated life is."[*]

.

Mathematicians are a separate species. Or at least a rare subspecies of *Homo sapiens*.[†]

.

And then there is, by contrast, the famous anecdote about that other great mathematician, Johann Carl Friedrich Gauss, being given busy work by his elementary school teacher: add up all the numbers from 1 to 100.

.

Gauss, like von Neumann, gave the correct answer almost immediately. In this case, however, it wasn't because of superhuman computational capabilities. The young Gauss realized that pair-wise addition from top and bottom of the number list would give identical intermediate sums: i.e., that $1 + 100 = 101$, $2 + 99 = 101$, $3 + 98 = 101$, etcetera, for a total sum of 50×101 which is equal to 5050—the result with which he startled his teacher a few seconds after the computational task was assigned.

.

So let us admit that mathematicians are a diverse subspecies. This is the reason that the nature of their creativity, like those of artists and scientists, is so hard to pin down. But even if mathematicians reside in ivory towers far removed from even those of scientists doing 'pure' science, their sometimes difficult to decipher dispatches out to the other domains of creativity have had an ever increasing influence on both science and art.

---

[*] I regret not finding him a suitable fit for one my cases studies, for he was 'a case' if there ever was one. His life was complicated, and it was complicated largely because of his freewheeling, hedonistic nature. He is often cited as an example of how Hungarians (noted for their mathematical genius) are actually an alien species.

[†] "Some humans are mathematicians; others aren't." (Jane Goodall in her book *In the Shadow of Man*)

THE BIOLUMINESCENCE OF ABSTRACTION

Question: What makes a glow worm glow? Answer: The emission of photons from a chemoluminescent reaction in which the pigment luciferin is oxidized by the enzyme luciferase. Question: What makes our hearts glow listening to the "Ode To Joy" in the last movement of Beethoven's Ninth Symphony? Answer: It echoes the complex four movement structure of the whole symphony, as a resonant mini-symphony embedded within the whole symphony, and the unexpected introduction of a choral section in a symphony taps into the part of our brains that is hard-wired to respond to the human voice.

.

Are these answers abstract or not? Are they meaningful to those who aren't chemists or musicologists? Are they complete answers?

.

The two most pure and abstract creative constructions of our species are mathematical and musical. Given that we have the innate perceptual and cognitive ability, and proper training, both math and music seem capable of consistently producing a primal biological experience of pleasure, as well as sometimes eliciting an apparently mystical understanding. How can this be, considering they are so abstract and apparently disconnected from reality?

.

In attempting to understand the paradox of abstraction often producing the most concrete and visceral reactions, one has to keep in mind that there are two levels of appreciation in this realm that reflect two levels of understanding. One I have to call 'superficial' because it is closer to the perceptual surface, but I do not mean to imply it is superficial in any pejorative sense; I certainly don't mean that it is necessarily uncomplicated or unsophisticated. The other I will call 'deep' but only because it is an understanding of relationships below the perceptual surface; it may sometimes be less profound than superficial understanding.

.

Most people have no musical training and no cognitive understanding of the abstract relationships from which all music is created. Yet the vast majority of people do have a profound response to music. You don't have to understand sonata allegro form to appreciate and respond to tuneful Mozart sonatas. You don't even have to understand what a 'tune' actually is. Newborn babies, even animals, show emotional responses to music. The goose bumps and tingling in our spine we can get from a sudden arpeggio, we still get even if we don't know a musical chord from a spinal cord. For those of us without extensive musical training, most of the underlying *deep* appreciation of music is not possible. Nonetheless, our *superficial* appreciation can be profound. If this weren't true, music wouldn't have such preeminence in virtually all cultures.

.

I have polled my students with musical training about whether their musical education has increased their aesthetic appreciation of music. Of course most do say it has, but some say it has had the opposite effect, that it has decreased their ability to respond emotionally because they are too aware of the underlying structure.[*] What is even more interesting is that when I ask all my students how much music means to them, often those rating it most important are not the musically educated. (Of course those that are actually musicians, not merely musically educated, consistently rank music as extremely important to their lives.) The point is that what I'm calling the superficial response often is profound, and certainly more common than deep responses.

.

However, I think that as a general principle, deep understanding is to be desired, for far more often than not it enriches one's appreciation. The composer and musicologist Dan Greenburg draws an analogy between baseball and music that I will shamelessly steal—and elaborate on.[†].

.

Imagine that a tear in space-time causes a member of the audience at an 18th Century Mozart concert to be deposited in the empty seat next to you in the stands of Yankee Stadium. Assume neither of you find this odd, so you watch the baseball game together. He will no doubt, at first, be fascinated by the goings-on, especially the athleticism: balls being thrown at amazing speeds, sometimes struck by big sticks wielded by big guys, who then run off toward some sack on the ground to their right, while whatever guy is near the flying sphere leaps into the air to catch it and fling it back toward the centre of the arena. He will no doubt also be totally confused at to why the players switch places regularly and why some things cause the

---

[*] My 'ear' isn't exactly sensitive. I couldn't tell a B-flat from a flat tire. I remember a friend with perfect pitch telling me my turntable wasn't spinning at the right speed, and it was driving him crazy because the music was out of tune. His remark broke the aesthetic trance the music had put me in.

[†] It is in one of his lectures for the Teaching Company about music appreciation.

spectators to cheer wildly. (An audience reaction to a ball getting so thoroughly clobbered that it flies over the fence surrounding the arena makes some kind of sense, but most of the crowd's responses would seem totally mysterious to our time traveler.) Before the ninth inning is over you notice your guest is getting a little antsy, and when the game is over he doesn't seem to understand why. You take him home and microwave dinner, which he seems to find even more interesting than his first encounter with the great sport of baseball. Now this is the World Series, and you have tickets for the next game, so you are very disappointed he doesn't seem to really want to go again. Rather than introduce him to the wonders of television, you spend the evening giving him a crash course in the rules of the game, and the next day take him back to Yankee Stadium. His understanding during this second game will be deeper than it was the day before. I don't know if I'd bet on the Yankees, but I would bet he'd appreciate his second experience of the game of baseball more than he did his first.

.

So it's the bottom of the ninth with the Yankees down by three runs. The bases are loaded, two outs, and a three-two count on the batter.* The pitcher winds up and sends a fastball down the centre of the plate. The batter, a real slugger, swings—and suddenly space-time twitches again, and you and your new friend are at a Mozart symphonic concert in 1786. Assume, in this fantasy, that you have never attended a classical music concert. You'll find the tunes charming. You'll find the 'athleticism' of all those people blowing horns or stroking strings impressive. But you won't really understand why there are four 'innings' to the 'game' or the why the tempo in the first is so fast, so slow in the second, the third sounds like a dance, and the fourth fast again with certain tunes repeated. So your friend takes you home and explains the structure of the classical symphony, and so when you join him for another performance the following night—well, you get the idea.

.

Even if it is fair to say that deep understanding below the perceptual and intuitive level almost always enhances appreciation, the extent to which it does so varies greatly depending on many, many factors. Is the history of the object of your contemplation important? How complex is the deep structure and how integral is it to 'getting it'.

---

* If the reader doesn't see the drama in this situation because of ignorance of the rules of baseball, I've already made my point.

How much of the pleasure is *embedded* in the cognitive understanding of the deep structure?

.

In the case of mathematics, the answer to these questions is that it is more important than in sport or art or even science. It is true that to fully appreciate—or be shocked by—the innovations of Beethoven on the Classical Period musical forms one needs to understand those forms, but the 'superficial' appreciation is still so intense that it could hardly be considered essential. You don't even need to know that symphonies have four movements and don't traditionally include choral works to be moved to tears by Beethoven's Ninth and intuitively understand when the 'game' is over. In science, generally it matters more to appreciate the deep structure of a theory or hypothesis, but here superficial metaphor can still serve. I'm told the mathematics beneath the surface of Einstein's General Theory of Relativity is not something I'm likely to ever grasp, but I still can appreciate it for its predictive powers—and even get some understanding of its elegance as a geometrical—rather than mystical—explanation through the metaphor of gravity being like a dip in a rubber sheet produced by the presence of a heavy object.

.

And then there is mathematics. The appreciation may be as profound as in art and science, but it is the most dependent on deep understanding, for there is so little at the perceptual surface. Often cited as the most "beautiful" equation of all time is $e^{i\pi} = -1$. This 'translates' to a statement that Euler's number when raised to the power of the square root of negative one times pi is equal to negative one. Got that? Are you moved? Well if you are then you are mathematically more sophisticated than me (and I assume the vast, vast majority of folk); for although this equation describes an inexplicably perfect and simple relationship between the major constants in mathematics (including the irrational number pi and the imaginary number i) that is a wonder to mathematicians, it offers no access to superficial appreciation, even through metaphor, for most of us.

.

So one almost always has to scratch the surface and go below to appreciate math as math. And it is a hard surface. The aesthetic appreciations down there are only available to those willing to spend many years of their lives mastering the tools required to break ground.

.

Fortunately indirect superficial (and profound) appreciation is possible in many cases, and is becoming more so, even as deep appreciation becomes more difficult. It occurs in the *application* of the math to other domains. Its importance in science is well established and continues to grow. What is a new and interesting trend is its increasing application in the arts. It is often based on *visualisation* of the relationship, what might be compared to a form of synesthesia. Everyone has seen the brilliant, fluid visualisations of music produced by running music through a computer that applies rules to assign shapes and colours to the ever changing frequencies and intensities of each sound component. This is the completely abstract deep structure of musical relationships translated into something superficially accessible through our vision. Normally the translation is into sound where the translation is infallibly accurate and consistent, for it is the translation that is the art itself, its *raison d'être*. Of course this is not true with the computer generated visual translations, where the rules are flexible and arbitrary. One can fiddle with the rules; e.g., one can choose to translate increased volume into changes in colour or brightness or intensity. There is no 'correct' translation. There are an infinite number of possible translations. Yet it would be incorrect to say that there isn't a relationship between the deep structure of the musical relationships and the superficial image translations, for there definitely is. In fact what is exciting about this is the increased variety of possible, equally valid translations. Variety is the spice of life—and the meat and potatoes of art.

.

Math in collusion with the agents of science and technology is opening new and exciting territory for artistic exploration. This is especially true in two related areas that are the subject of the next two sections.

SMART OXYMORONS: RANDOM ORDER AND CHAOTIC STRUCTURE

Pour a cup of sugar in a large pitcher of water. And wait. At first most of the sugar granules will rest at the bottom, but eventually they will dissolve and apparently disappear as they go into solution. Initially a sample of this solution near the bottom of the pitcher will be sweeter than one nearer the surface, but eventually the sugar will he equally dispersed throughout the solution. The important philosophical question about this transition is whether it is a movement toward or away from 'order'. One might argue that it is more *orderly* to have sugar separated from the water. On the other hand, one might argue that the homogeneity of the mixture of sugar and water that eventually occurs is more *orderly*. If you inclined toward the latter viewpoint, you must consider the eventual heat-death of the universe, where the logic of randomization means all particles of the once complexly structured and heterogeneous universe will be distributed equally in space for static eternity, as perfect ultimate order. To make it personal: imagine that you hire some librarian (having this interpretation of order) claiming to be a master at putting anyone's book collection 'in order', and he goes into what you believe to be an already moderately well-organized library and proceeds to shred all your books into tiny pieces and then carefully spreads the tiny shreds, nicely equidistant from each other, on the floor. And asks to be paid for this accomplishment.

The Second Law of Thermodynamics predicts such an eventual grand shredding and uniform spreading of *everything*. Does this mean we are progressing toward the ultimate order? When we use the word 'order' do we really mean simplicity and uniformity? And what about the obvious movement toward complexity and diversity that is the result of evolution? Does that refute the Second Law Of Thermodynamics, and docs it mean we arc actually moving toward more disorder?

It has been argued by creationists (who would like to claim that physical science refutes biological science[*]) that evolution is contrary

---

[*] It is interesting and amusing to see those opposed to science claiming science refutes science. This has become more and more common since the everyday efficacy of science and technology is so obvious. Most of the so-called 'refutations' of evolutionary theory I've seen or read rely heavily on misinterpretations of science to 'support' their case.

to the Second Law of Thermodynamics—which is often called the Entropic Principle. They do this with their distorted version of the Anthropic Principle, which in their wishful-thinking version posits a Grand Designer busy making things more complicated but more 'orderly' in terms of a certainly far from uniform and homogenous Grand Design.*

And if 'order' is an ambiguous word, what about 'chaos'? According to Biblical and other scriptures the world, the universe, was once chaotic. And then some Supreme Being came along and imposed order on the chaos. In Greek mythology, Chaos (Khaos) was the primeval state of the universe from which the first gods emerged. It was a homogenous mixture of what the Greeks considered to be the four elements (earth, air, water, and fire). In what presumably was equivalent to a Big Bang their gods spontaneously burst forth from this uniform soup, and the world as we (they) knew it came into existence. The ancient Greeks considered the initial state complete *disorder* and what followed the birth of their gods the beginning of order. And of course the Bible (in *Genesis*) describes the world (i.e., the universe) being "without form or void" until God put in the ultimate six-day work week putting things in order.

So is a shelved library more orderly than its shredded remains? Or is order equivalent to chaos? Is the end of the universe like the beginning: homogenous, chaotic, "without form or void", *and* thus perfectly 'ordered'?

Finally, but inextricably intertwined with the ideas of order and chaos, is the idea of 'freedom'. Were I here to introduce the eternal conundrum of free-will, I'd moving into philosophical realms that are outside of the purview of this book. However, there is a concept called "degrees of freedom", and there are serious questions about a new view of causal determinism that are relevant.

---

* The Anthropic Principle actually is the simple statement that humans being around to observe their universe may be highly *improbable*, but that we *are* here obviously proves it is *possible*. It is very improbable that one could correctly guess a coin toss correctly even 10 times in a row (.5^10 or just 1 time out of 1024), but it is very likely that someone in a line of 2000 individuals will—and if it is the first person in line who does this, that is no less likely than it being the last person. It also is no reason to think the game was rigged—and no reason for that person to think he is somehow special and the result of some grand design or designer or game rigger.

The phrase "degrees of freedom" has slightly different meanings in statistics and physics and engineering, but its core meaning has to do with the number of parameters that affect an outcome. If A and only A always causes B, there is no 'freedom' in the relationship. But as more and more variables, and their relative strength, can affect an outcome, the degree of 'freedom' in the system increases. It seems there are more degrees of freedom in even the simplest system than previously suspected.

.

It goes by several names: Chaos Theory is probably the most common. When a clock-work universe was the working metaphor and model in science, causal relationships were assumed to be linear. That is to say that the accepted schema went something like this: A (the cause) affected B, which affected C, which in turn affected D (the outcome). In such schemas no effect is greater than any other, and the whole process could be compared to the collapse of a row of dominos that leads to the final domino falling. What Chaos Theory introduced—or at least drew attention to—was the idea of thresholds beyond which even the slightest, most minute, change could become drastically amplified and produce totally unexpected results.

.

This extreme sensitivity to the tiniest differences in initial conditions has become known as "The Butterfly Effect". In 1972 the mathematician and meteorologist, Edward Lorenz, presented a paper at a conference of the American Association for the Advancement of Science entitled "Predictability: Does the Flap of a Butterfly's Wings in Brazil set off a Tornado in Texas?" The answer he gave was a resounding yes. Researchers in many diverse fields have found the underlying mathematical principles he explicated can explain phenomena which previously had been considered so chaotic and unpredictable that they were written off as so complex as to be effectively random. Chaos Theory illuminates much in evolutionary theory, economics, physics, sociology, neurology, and ecology—just to name a few fields.

.

Its relevance to art results from it being accessible at what I've been calling the superficial level, and this access is through visualization that is made possible because of advancements in computer technology. But before turning to this important topic, I want to focus on the philosophical, scientific, and artistic implications of these changing conceptions of the nature of causality, order, randomness, and chaos. These ideas, often based on arcane mathematics, have become, in one form or another, a powerful influence on the way artists and scientists think.

.

The clock-work universe model is disheartening because it doesn't seem to allow any space for any kind of freedom and flexibility, but at the same time it is reassuring. A simplistic determinism is easy to deal with. There is something comforting about the orderliness of linear cause and effect, for the same reason that simple 'solutions' to problems with complex causes are so readily and unthinkingly embraced in social policy decisions.* The major disturbing insight of Chaos Theory is that small changes in what would be considered insignificant things—things beyond our practical control—can be the root cause of huge positive or catastrophically negative change. (Damn butterfly! Who'da thunk he'd cause a tornado. How can we control such things?) There are far more degrees of freedom—which can be considered for all practical purposes random—in complex deterministic systems than dreamt of in our previous philosophies.

.

Long before Chaos Theory, science fiction writers dealing with time travel played with this idea. What if a time traveller went back in time and—while conscientiously trying not to influence anything that could have any repercussions on the future—caught the attention of a fellow walking down the street who thought this stranger oddly dressed, and this curious fellow paused to stare for a few moments, causing him to miss his train home. Arriving home later than usual, his wife is annoyed, so they don't have sexual intercourse, which would have happened had he been home on time for a change. And so a child that would have been conceived that night isn't—and a whole ever expanding generation of offspring do not occur. Perhaps one of those offspring would have been Adolph Hitler—or Albert Einstein, or you, or me.

.

---

* I call this the "Panacea Syndrome". Social engineers do a lot of harm because of the simplistic reasoning that leads them to believe an evil has a single cause that if removed will fix everything. That Mister Jones was shot in the head by his wife isn't *just* because we don't have stricter gun registration, and in fact it may not have been a factor at all in his death. If there wasn't a gun in the house, his wife might have poisoned or stabbed him. And if he hadn't missed his bus because his shoelace broke, he wouldn't have been given a ride home by his secretary which then led to an affair with the attractive young woman, which his wife found out about and which pushed her over the edge of tolerance for his philandering. Stricter quality control of shoelaces wouldn't have saved him. (This is not to say that stricter gun control isn't needed., but there are many resaons for that.)

Only good can come from recognizing that even such fundamental concepts as causality and order are far more complicated than previously assumed. Our world becomes more or less orderly (depending on your definition) through evolutionary natural selection. The universe as a whole moves toward greater or lesser order (depending on your definition) because of the Entropic Principle. Random events, while perhaps not truly random in the sense of quantum physics' elementary particle decay, are seen as much more important in science and art than ever before. The potential importance of what was once considered trivial is recognized.

These ideas are causing a paradigm shift in the arts, as well as the sciences. What has brought them to the forefront is technology: the computer and the computer monitor.

IMAGINARY NUMBERS: MONSTERS IN A FRACTAL SEA

Imagine something that is impossible. And if you succeed in doing so, then try to understand how it is possible that this impossible thing is having a profound influence on what actually is.

.

The Pythagoreans loved numbers, whole numbers that is, for they felt they were real and rational. Numbers that could not be expressed as the relationship of two integers (e.g., a fraction) were considered 'irrational', a term that is still used. Pythagoras considered irrational numbers as heresy and when one of his disciples, Hippasus of Metapontum, tried to present a proof of their existence, Pythagoras had him killed—or at least so goes the legend.

.

The square root of 2 is an irrational number, but while it can never be defined with absolute precision, the more decimal places you take it to, the closer squaring it will approach the integer 2. (1.4 times 1.4 gives 1.96; 1.414 times 1.414 gives 1.999396; etc.) Unlike Pythagoras, most of us can live with that, and if we call the square root of 2 'irrational', we still accept it as 'real'*.

.

But what about the square root of negative one? All negative numbers when multiplied by themselves give a positive number. So there can be no real number that when squared yields a negative result. Such a number must be imaginary, and why on earth would anyone be interested in it? You can't even conceive of it. Descartes back in the 17th century coined the term, and like the term Impressionism in painting it was originally both descriptive and pejorative.

.

But damned if these imaginary creatures didn't prove useful. Mathematicians use the letter i to indicate the square root of negative one, and this number figures prominently not just in theoretical mathematics but also in real-world applied math. That imaginary numbers might find the weird world of quantum mechanics a congenial environment isn't too surprising, but they are also very important in the practical field of electrical engineering where, for example, in doing the math on AC voltages using them, a mistake could have the very real world effect of electrocuting someone.

---

* 'Real' is the real term mathematicians have adopted for the larger set of numbers that include irrational numbers, but does not include 'imaginary' numbers.

Where this amazing product of the mathematical imagination has had the most effect on art is in the realm of fractals. Just as what just might be an imaginary construct, that of the 'subconscious', has resulted in the ubiquity of surrealist images, so too have imaginary numbers resulted in the ubiquity of fractal images. They are everywhere, and their deep structure remains a mystery to most of even their most enthusiastic admirers.

Many people don't even realize that these beautiful fractal images are nothing more than mathematical *graphs*—no different in kind from a pie chart or bar graph or a grid with a line indicating some trend over time. *A fractal image is just a graph of an iterative process applied to complex numbers.* Two of the terms in this definition need to be clarified.

Iteration simply refers to doing something over and over again, but in this context implies performing an arithmetic procedure on the result of the previous application of that arithmetic procedure. For example, double a number and then double the product repeatedly. (2 times 2 equals 4. 4 times 2 equals 8. 8 times 2 equals 16, etcetera. This leads to the sequence of 2, 4, 8, 16, 32, 64, 128, 256—which explains the importance of these specific numbers in computers which are binary in nature.[*]) Another example is that if you apply to two numbers the iterative rule of adding them together and then adding the sum to the last number what you generate is called a Fibonacci sequence. (Start with 0 and 1. 0+1=1. 1+1= 2. 1+2=3. 2+3=5. 3+5=8. 5+8=13. Continue on and you get the sequence 0, 1, 1, 2, 3, 5, 8, 13, 21, 34, 55, 89, 144, 233, 377, 610, 987, 1597, 2584, 4181, 6765, 10946, 17711, 28657, 46368, 75025, 121393, 196418, 317811, etc. This sequence, generated by what seems a totally arbitrary rule, amazingly occurs repeatedly in nature, from the distribution of seeds on sunflowers to the spirals on shells. This is another example of how math seems to lie beneath the surface of everything. It's enough to make one believe God is a mathematician.

A *complex number* is a number containing a 'real' part and an 'imaginary' part. For example, 4+i is a complex number for 4 is a real

---

[*] Each bit (which is binary) in a traditional eight-bit byte doubles the information that can be stored. This explains why early home computers (such as the Commodor 64) could only display sixteen colours if they were just using half a precious byte of memory to store the colour. (Using the other half of the byte would've allowed the display of 256 colours.)

number and i is the imaginary number, the square root of -1. One can perform some conventional arithmetic operations on complex numbers by treating the real and imaginary parts separately. For example, (3+2i) + (4-5i) = (3+4) + (2-5)i which equals the complex number 7-3i.

.

What does all this have to do with those awesome images of fractals? Well, before attempting to answer that question, one defining characteristic of fractals has to be considered. Fractals display the characteristic called self-similarity. This means that repeating the iteration produces results that bear *some* similarity to the previous result of the iteration—something that shouldn't be too intuitively surprising. The usual metaphor is one of magnification: each iteration is considered a magnification of the previous one. The traditional comparison is looking closer and closer at the shoreline of Great Britain. There is great similarity between a close-up of a detail of the shoreline and the view from a satellite: jagged coves, etc. And then there is the question as to how long is that shoreline—really? Well, it depends. It depends on how much you zoom in. Before zooming in much, you are rounding, ignoring the smaller bays and indentations. But as you zoom in and take these into consideration, the length progressively increases. So there is no absolute answer to the question. It depends on the degree of magnification. Fractals, like irrational numbers, are not integer-based. The concept of dimension has always been considered integer based—but fractals break from that tradition. Something is not either two or three dimensional; it can have a fractal dimension in between integers. For example, ~1.58, is the fractal dimension of the fractal called the Sierpinski Triangle.

.

Finally and most important in understanding the nature and profound implications of fractals is that fractal images are nothing more than graphs of different levels of iteration according to some set of *arbitrary* rules. These rules include assigning colours and the three parameters of colour (hue, saturation and brightness) to the results of the calculations. These rules are as arbitrary—and as easily changed—as the rules that are used to 'translate' music into visual images. Some rules, when applied, produce startlingly beautiful images (graphs), but then some rules don't.* The creation of beautiful

---

* When I began using fractals in my own visual art, an acquaintance asked what computer program I was using. I recommended *Fractint*, probably still the best and most flexible fractal generator around. She was profoundly disappointed that it didn't automatically produce images worthy of being considered artistic.

fractal images is an art—albeit an art rooted deeply in math. It is also rooted in scientific technology, for this whole new territory would never have been explored were it not for the computer revolution—the subject to be considered next.

.

One of the most intriguing characteristics of fractals is the complexity they display within their self-similarity. Anyone who has played with a fractal generator on their computer knows this. Each 'magnification' pleases, just as variations on a musical theme do, reassuring with familiarity and refreshing with novelty. Their potential isn't limited to the visual arts; they are being used in the composition of music. Fractal Geometry, like the related field of Chaos Theory, has degrees of freedom that are virtually infinite. These math-based innovations may well become as influential in the future as the Golden Ratio and single-point perspective were in the past.

CASE STUDIES: BENOIT MANDELBROT, M.C. ESCHER

The "Father of Fractal Geometry", Benoit Mandelbrot, was born in Warsaw in 1924 to an intellectual Lithuanian-Jewish family. His father was in the clothing business and his mother was a medical doctor. One of his uncles was a renowned Parisian mathematician, Szolem Mandelbrot, who nurtured Benoit's interest in mathematics. As a boy, he was fascinated by the rigid, deterministic game of chess and was considered a prodigy, but he gave it up ("retired" he says) at the age of eleven, shortly after his family fled from Poland to France in response to the rise of anti-Semitism. He says he lost interest in the game because he couldn't find any real competition in his new homeland.

.

Mandelbrot attended the Parisian Lycée Rolin then the École Polytechnique*, where his geometric approach to the problems on the entrance exam was considered bizarre, albeit successful—and foreshadowed his subsequent visual and geometric approach to real life problems in the sciences. He continued his studies at the California Institute of Technology (Caltech), where he specialized in aerodynamics, specifically studying questions about the nature and unpredictably of turbulence—again something that foreshadowed his important discoveries about chaos theory and fractals. After several years at Caltech, he returned to Paris, obtaining a Ph.D. in Mathematical Sciences from the University of Paris in 1952, while working at the Centre National de la Recherche Scientifique—with a year off at the Institute for Advanced Study at Princeton, sponsored by that other giant of mathematics, John von Neumann. In 1958 Mandelbrot took up permanent residence in the United States.

.

It was in 1982 that he published the modern classic *The Fractal Geometry Of Nature*. This is arguably one of the mathematical works most widely influential across disciplines and domains of creativity to be published in the 20th century. Precisely why this is so has a lot to do with both synchronicity and the coming together of art and science.

.

---

* One of his teachers was Gaston Maurice Julia, who devised the formula for the "Julia Set" which has a direct relationship to the "Mandelbrot Set". In fact, Julia had sunk into relative obscurity until Mandelbrot began to achieve fame and recognition for his related work on the Mandelbrot fractals.

1982 was also the year that the first truly functional, graphical home computer was released: the Commodore 64. It was also the year I (and millions of other people) bought it. It was the following year that I (and many, many other people) also bought his book *The Fractal Geometry Of Nature*. It was also around then that I wrote a simple program, using the C64's very basic BASIC, to generate images based on one of the simple formulae at the heart of fractal generation. More sophisticated "Fractal Generator" software also was already appearing for the C64, based on Mandelbrot's ideas. In those early, heady years, Mandelbrot must have been traveling a lot, for he even gave a lecture in a small city 100 kilometres away from my Northern Ontario home. I couldn't go, but a friend of mine did, and it was all he could talk about for months afterwards.

.

Why all the excitement? Well, as I hope I've explained, fractal images are really nothing more than *graphs* of mathematical functions, something one doesn't normally get excited about. Ah, but they are the most beautiful of graphs, easily appreciated for their pure aesthetic qualities by people who are mathematically illiterate (innumerate) and appreciated even more by those who can at least partially perceive their origin in the sublime and misty regions of mathematical abstraction. They are produced by such extremely simple mathematical functions and yet are so complex.

.

But the important thing here is that they are colourful visual manifestations of subtle mathematical functions, only made possible by the computing power to iterate the simple formulae on which they are based and present the results in graphic colour. Fractals would have never captured the popular imagination—and probably not even the scientific community's imagination—were it not for this happy synchronicity of advances in computer science and Mandelbrot's eclectic interest in "mathematical monsters".

.

The history of fractals goes back to the 17th century and the philosopher and mathematician Leibniz (co-discoverer, with Newton, of the principles of calculus). It was in 1872 that the first graph that could be considered fractal was produced by Karl Weierstrass. Other classic fractal graphs appeared in the following decades, including the Koch snowflake and the Sierpinski triangle and carpet. However most mathematicians still considered fractals "monsters", just as Pythagoras considered irrational numbers monstrous. (Although, unlike Pythagoras allegedly did, no one put a contract hit on a colleague interested in these "monsters".)

Historically, to physically graph the results of the successive iterations that define fractals was far too labour intensive to attract much interest, even among those who found these monsters interesting. Then along came the computer revolution. Mandelbrot was the right man in the right place at the right time. With the now possible computer-assisted visualization of the deep structure of this new field of mathematics, fractals captured the attention of not just mathematicians, but also the general public in love with the potential of the new computer technology they could put on their own desktops.

I believe this technology is one of the most important forces in the future development, evolution, of creativity. That is the subject of the next section.

Mandelbrot has a small planetoid named in his honour. He retired in 2005 from his faculty position at Yale and presumably is observing with great pleasure the many fruits of his creative labours.

Another man, an artist, working before the computer revolution, but with an intuitive grasp of the importance of mathematics in the arts is M.C. Escher. Prints of his artworks are as popular and ubiquitous as those of fractals and of the major surrealists.

Maurits Cornelis Escher was born in Leeuwarden (Friesland) in the Netherlands on June 17, 1898. Again fairly typically of creative people in their youth, he was not particularly successful academically, but what is of particular interest, since his art works were to eventually be so inspired by mathematics, is that he was to write of this time that "at high school in Arnhem, I was extremely poor at arithmetic and algebra because I had, and still have, great difficulty with the abstractions of numbers and letters. When, later, in stereometry*, an appeal was made to my imagination, it went a bit better, but in school I never excelled in that subject. But our path through life can take strange turns."†

---

* A term for a branch of geometry.

† S. Strauss, "M C Escher" (*The Globe and Mail*, 9 May 1996).

At the age of twenty-one he attended the Haarlem School of Architecture and Decorative Arts where he switched from studying what his father wished, architectural design, and concentrated on the so-called 'decorative arts'. In 1922 Escher, now fluent in drawing and making woodcuts, travelled to Italy and Spain. There the architecture of Italy and the mosaic tiling at the Alhambra*, the 14th century Moorish castle in Granada Spain, inspired his earliest important works. It was in Italy that he met Jetta Umiker (also an artist), whom he married in 1924 and with whom he was to have three children, and by accounts live a fairly conventional bourgeois family life.

Mauk (as his family and friends called him) and Jetta settled in a suburb of Rome, but travelled frequently throughout Italy, with both he and his wife sketching the landscape and local architecture. When the Fascists began their rise to power in 1935, Escher moved his family to Switzerland. The following year they took an excursion voyage on the Mediterranean, which included his second visit to Alhambra which forever confirmed and reinforced his obsession with complex symmetry and 'filling the plane'. From this point on in his creative life geometrical mathematics and art were to remain inextricably entangled.

Unimpressed with Switzerland, Escher and his family moved to Belgium where they were to reside until 1941 when again they were forced to move because of politics: this time the invasion of the Nazi Germans. They fled to Baarn in Holland†, the country where they were to reside until his death.

It is one of those common apparent contradictions in the lives of creative individuals that what they profess to being most 'unprofessional' in their understanding of is precisely what they end of up being most noted for. Escher remarks that "By keenly confronting the enigmas that surround us, and by considering and analysing the observations that I have made, I ended up in the domain of mathematics, Although I am absolutely without training in the exact sciences, I often seem to have more in common with

---

* These famous mosaic tiles have been the subject of numerous mathematical analyses ever since.

† The Netherlands, of course, also fell under Nazi occupation, but Escher and his family had no where else to flee.

mathematicians than with my fellow artists."* He considered it an "open question" whether his work "pertains to the realm of mathematics or art." I would say the answer is *both* realms, for Escher, despite his modesty regarding his mathematical sophistication, was working down in the deep structure of mathematics and not merely using math as tool.

.

One piece of evidence of this deep relationship is that he published papers which led to his being considered a research mathematician by mathematicians: he was honoured by peer review and recognition.†
Another piece of evidence is his relationship with Lionel and Roger Penrose, father and son and both noted mathematicians—Roger also a physicist and his father a medical geneticist. Roger Penrose, upon seeing one of Escher's 'impossible figures' became fascinated with the concept and sent Escher an idea for a new impossible figure that was to be incorporated in his famous works: *Waterfall*. This work is based on Penrose's now famous 'impossible triangle' and is a picture of water flowing downstream to a waterfall turning a waterwheel, at the base of which is the source of the water stream! It is a sort of perpetual motion machine. It is also the visual equivalent of the Shepard Scale in music where a sequence of tones appears to continuously rise in pitch forever, but really is a closed loop.

.

Escher's major creations can be sorted based on five general categories of artistic/mathematical exploration.‡

- Tessellations (or tilings) and polyhedra.
  - o E.g., "Regular Division of the Plane with Birds" and "Order and Chaos"
- Images based on impossible figures and perceptual distortion
  - o E.g., "Waterfall" and "Relativity"
- Explorations of the nature and logic of space
  - o E.g., "Circle Limit III" and "Mobius Strip II"
- Morphings
  - o E.g., "Metamorphosis III" and "Fish and Sky"

---

* Quoted in E Maor's *To Infinity and Beyond* (Princeton 1991).
† In his first paper, published in 1941 and called "Regular Division of the Plane with Asymmetric Congruent Polygons", he explored colour-based geometric division and proposed a system of categorizing shapes, colours and symmetrical properties. He was to publish other mathematical papers that are still considered important and seminal.
‡ Images of the works cited as examples can be easily found on the Internet.

- Self-referential works
  - E.g., "Drawing Hands" and "Three Spheres II"

This is a very rough taxonomy and most of his works can be placed within more than one category. What is worth noting is that all explore the deep structure of math and logic by making it visual—and concrete. For example, in his Mobius Strip he draws ants marching forever along the single surface of this two-dimensional object, something which adds deep aesthetic and philosophical resonance to this commonplace but mathematically interesting object which anyone can construct by first twisting and then gluing together the ends of a strip of paper.[*]

By the 1950's Escher's reputation and fame was such that he supplemented the income from his popular prints with numerous lectures delivered to very diverse audiences. His appeal was across all the domains of math, science and art; and in each realm his creations were appreciated at both what I've called the superficial, as well as the deep level. It isn't surprising that Douglas Hofstadter, in one of the most multi-disciplinary philosophical books published in recent decades, made Escher a key figure in his discussion. I'm referring of course to the modern classic, *Gödel, Escher, Bach: an Eternal Golden Braid.*

In 1969 Escher had to cancel a lecture tour because of deteriorating health, but he finished his last work before his death three years later—when the modern world lost one of the artists most influential in bridging the gap between the creative domains of math and art. This final work is a woodcut called *Snakes.* It consists of recurring patterns that fade to infinity, both to the center and the edge of a circle, with snakes transversing the circle and the patterns in it—with their heads sticking out of the circle. Are these snakes a reference to the serpent that tempted Man to take a bite of the apple of knowledge and thus spend eternity and infinity immersed in patterns?

---

[*] Something about this image reminds me of the myth of Sisyphus popular with existentialist thinkers dwelling on life's ultimate futility. Sisyphus spends eternity pushing a rock up a mountain only to have it roll down again, so he has to start heaving it up again. Escher doesn't have people, only ants, marching along on this eternal mobius strip, but somehow that just makes the idea more subtle and poignant. One can associate a lot of other things with his image (e.g., Nietzsche's concept of eternal recurrence), but of course that is what the best art does: subtly resonant with many, many other relationships.

THE SHOCK OF GETTING WIRED

"Computing is not about computers any more. It is about living."
—Nicholas Negroponte (*Being Digital*)

"In this electronic age we see ourselves being translated more and more into the form of information, moving toward the technological extension of consciousness."
—Marshall McLuhan (*Understanding Media*)

The year Escher created his last work, scientists created ARPANET, the original Internet. And in the five years after Escher's death, Steve Wozniak and Steve Jobs co-founded Apple Computers, Bill Gates and Paul Allen founded Microsoft, and the first home computers were invented. The mid-seventies was a time of fierce competition for this new market, with the Apple I and II, the TRS-80 and the Commodore Pet being the major contenders.* Although he wasn't singing about computers, Bob Dylan had it right: Times, they were a-changin'. The social effects of these two technological advances, first the personal computer and then the Internet, almost make the effects of the political radicalism that had just preceded them seem minor by comparison.

The effects of technological change had already become *the* hot topic of investigation for cultural historians. The University of Toronto professor Marshall McLuhan had risen to the status of oracle regarding this topic—and pop culture in general. This was true despite the fact he was a devout Catholic, very unhip, and almost a parody of the Ivory Tower academic. Furthermore, like the Delphic Oracle he often spoke in riddles and made apparently contradictory pronouncements. Unfortunately McLuhan suffered a stroke around the time the computer revolution was really getting started and never had the opportunity to witness and pontificate on its effects.

What McLuhan and many others realized was that technology was more responsible for change in social institutions and culture than

---

* I remember hemming and hawing about which of these to buy—and then keeping my purse strings tightly closed until the Commodore 64 came out a few years later.

any other single factor. And changes in social institutions and culture are the major factor responsible for changes in art and catalyzing creative revolutions. Behind all this is the science that made the technology possible. Ergo, it could be argued that science is the ultimate culprit, the secret agent that has always been undermining the status quo in both everyday life and art.

.

To cite two obvious examples, the printing press forever changed communication, and the automobile forever changed transportation—and both rearranged social structures profoundly. But before the computer revolution, the prognosticators, the futurists, the science fiction writers, all seemed to favour another momentous break-through in transportation, not communication, as the most likely next catalyst to social change. For example, space travel, not a global communication network, was, or so they thought, in the cards. Few thought some machine that could do arithmetic at a fast rate, a sort of glorified calculator, would be of interest to anyone, and would soon become as common—and be considered almost as essential to the average citizen of an industrialized country—as possession of a car. Who would have thought this mere gadget would be the thing to irrevocably change civilization through its effects on art and science—and everyday life? Take a minute and think about this.

ANALOG MADE DIGITAL: INTEGERS FREED AT LAST

Take another minute or two. Watch a watch with hands. Watch a watch with numbers. The difference you will see is one of the most fundamental distinctions of all. It is the difference between integers and real numbers, between atoms and waves, between the discrete and the continuous, between determinism and indeterminism, between absolute and relative truth (and morality), between fading and oblivion, *between digital and analog.*
.

The word 'digital' has acquired a variety of denotations and connotations. It is often used to refer to anything computer-related—even though there are analog computers. Again it is time for the naming of parts.
.

The current usage of these two terms refers to the important distinction between a continuum with an infinite number of intermediate conditions and a finite set of possibilities with no intermediate conditions possible. If the answer to the prosecuting attorney's question has to be a 'yes' or a 'no' with 'maybe' or 'sometimes' or 'sort of' being impermissible answers, you have a digital question. The fundamental philosophical principle proposed by Parmenides that something either exists or else it doesn't—that is a digital conception. Schrodinger's Cat, whose existence is indeterminate until the box in which it is interned is opened, is not digital. Mr. Pussycat only becomes digital when the box is opened at which point his luck holds out or gives out. It is interesting that the idea of quanta in physics has come to be mistranslated in common parlance to mean huge, as in a 'quantum leap in understanding', when it actually means small but discrete, with no intermediate state such as when an electron jumps to a new orbit.
.

If you and your friend are asked what time it is and you both look at an old-fashioned analog watch with minute and second hands, your answers will be estimates and subjective and may differ slightly. If you both look at a digital watch you will both give exactly the same answer, but in one sense it too will be an estimate, however one based on some arbitrary rule hardwired in the circuitry for when to move from one integer to another.
.

Digital information is often construed as being more precise and accurate, but actually it is less so. It is just more consistent. If something that is actually continuous is converted to digital information that means information is lost. If one converts, rounds, pi to the integer 3, one doesn't get increased accuracy. Traditional phonographic records are analog, and so inevitably when they are digitized ('digitally remastered') information is lost—something collectors of vinyl never tire of pointing out to people with CD—or MP3—collections.

.

The power of digital technology is in its clarity, in its lack of ambiguity, and in its conformity with formal logic and absolutism. It is something to warm the hearts of any surviving integer-loving Pythagoreans. And the philosophical implications of this technology becoming dominant are important. It affects the way we perceive the world because it seems to quantify the qualitative.

.

If one thinks of memory as the storage of information, an interesting trade-off is involved with the transition from analog to digital. As a general principle, it is reasonable to say that analog information is more detailed and accurate, but digital information is more robust and indestructible. A painting is analog. It will fade with time. A digital photograph of that painting will lose information present in the original, how much depending on the number of pixels and the number of bits devoted to describing the colour of each pixel, *but it is impervious to change over time.* And it, albeit a degraded version of the painting, can be duplicated endlessly without any further loss of information. When people copied video tapes of their favourite movies and passed them on to their friends, who then made copies and passed them on to their friends, each new 'generation' of the movie was of lower quality than the previous. When people now copy movies in '.avi' or any digital format the copy is a complete clone of the original. This may seem obvious, but it has profound implications.

.

The world is full of ever increasing copies, backups, of all of our creations. But many of these backups, at least the ones derived from analog creations, are inferior to the original. The astronomical prices rich people pay for original paintings by the great artists is evidence of the value we place on the uncompromised, undigitized, work. Still the fact we less wealthy folk can print digitized versions that *almost* capture the full glory of the original is nothing short of wonderful. And we, unlike even the richest man at a Sotheby's Art Auction, can possess thousands of masterpieces.

.

There is a danger in such digital wealth. The common remark about those with great material wealth also applies to all of us with great digital wealth: "He's a slave to his possessions."

SLAVERY REINTRODUCED: THE MICROCOMPUTER

We love our computers—when they work. We hate our computers when they don't. We depend on them, but we hate being dependent on them. Still they do allow us to be more independent of many, many things—except them. They are our slaves but we pay for them. And they, like wage-slaves in an eternal union deadlock with management, 'work to rule'. They may be our slaves, but many of us almost slavishly worship them. Computers and *Homo sapiens* have a symbiotic relationship. They couldn't exist at all without us, but now many of us find existence without them almost unthinkable. Not surprisingly, then, there is a lot of ambivalence in our feelings about computers.

Once I was on an absurdly over-crowded train in Romania, standing in the aisle amidst a crush of bodies and trying to get some air through an open window, when a young soldier pressed up next to me and whispered in English: "Be careful, there are thieves among us." We are in an over-crowded world of computer technology trying to catch our breath, and many people around us are warning that there are thieves among us. I'm not talking about Internet identity theft or phishing sites. I'm talking about the concerns many people have that computers are stealing our humanity.

It is facile—and I admit that I sometimes do it—to label as a Luddite anyone dubious about the positive effects of the computer revolution. The original Luddites were textile workers who violently protested the introduction of textile machines during the Industrial Revolution. They saw the introduction of technology that could do what they were doing as a threat to their livelihood and dignity—which, of course, it was. They would conduct raids to trash the new textile machinery. As far as I know there are no modern day Luddites who are breaking into offices and smashing computers, but there are people who, for whatever reason, do try to bring down the Internet or at least the intranets of major corporate or governmental organizations.* And there are a lot of people who do feel computers are a threat to their livelihood and especially to their dignity. Of course many are only trying to undermine evil applications of computers and technology.

---

* Ironically, these people are experts in the technology they are attacking.

The parallels to the Romantic Movement are obvious, as are the logical contradictions inherent in opposing advances in science and technology because they are seen as dehumanizing. The Romantics were allegedly the radicals, not the reactionaries. They were rebelling against the classical constraints on their art and the social constraints on their behaviour. But their hostility toward advances in scientific understanding and technology is indisputably ultra-conservative—a fear and hatred of the new if it is outside of their personal domain.

.

In recent times the humanist Neil Postman is a good example of a modern 'Luddite' who is seen by some as radical and by others as reactionary, when he is actually a bit of both. His book (coauthored with Charles Weingartner) *Teaching as a Subversive Activity* (1969) is, as the title indicates, a radical tract about educational reform. But a decade later he wrote *Teaching as a Conserving Activity*—which foreshadowed his change of focus. In 1985 he wrote *Amusing Ourselves to Death: Public Discourse in the Age of Show Business*, which is a 'shooting fish in a barrel' screed about the obvious idiocy of most TV and mass media entertainment made possible by technology. Then in his 1992 book, *Technopoly: The Surrender of Culture to Technology*, he pulled all the stops. The title says it all: technology is the enemy of culture. Postman, like most intelligent critics of technology, makes important observations about the dangers inherent in unthinkingly embracing all technological innovation as cultural progress. Nevertheless, he is wrong, even silly, in most of his analyses.*

.

Where he *is* right is where he says, as he does repeatedly, that we must learn to "use technology rather than be used by it."† This may seem to imply the absurd anthropomorphizing proposition that technology has sentience, and it can 'use' us to further its own aims— whatever they might be! However, I do believe what Postman is warning against is something more real: letting what technology *allows* us to do become what we feel we *must* do. It like the warning against letting our possessions run—and ruin—our lives.

.

---

* For example, he even questions the value of the invention of the printing press! He admits it fostered the idea of individuality, but claims this destroyed what he absurdly considers the sense of community that preceded it. He even claims it was to blame for the demise of poetry as a popular art form—obviously ignorant of its popularity and importance for hundreds of years after Gutenberg and only made possible by its invention and the resulting increase in literacy.

† Quoted from a 1996 PBS interview with Paige MacLean.

I'll repeat myself. Technology is a tool. That is its definition. A hammer is a tool. To a man with a hammer sometimes everything seems a nail. It can be used to pound in nails or pound in your spouse's head. A hammer is morally neutral.

.

Computers also are a tool. You can use them to hammer into shape a book you are writing or to hammer the email in-boxes of millions of innocent Internet users with spam offering to increase their penis or breast sizes.*

.

However, computers are very special tools because they aren't as specialized as most tools are—such as hammers, which are really only good for one thing. Computers have proven to be incredibly versatile.

.

They also differ from most tools in that they will do exactly what you tell them to do, without relying on you to guide them. I've known hammers that didn't hammer the nail I wanted them to, but instead hammered the nail on my index finger.

.

I initially referred to computers as slaves. I think this is a very reasonable comparison. The abolition of the moral abomination of human slavery created a vacuum, for we—or almost all of us—want something to unquestioningly obey our every command. We can't expect that of our employees, our spouses, our children—or even our pets. If we do naively expect it, we're bound to be disappointed.

.

My computer, on the other hand, is a real 'yes man'. If I say "speak", it speaks. If I say "speak faster" it speaks faster. It can even be programmed to automatically respond to a 'speak command' with the question "How fast, sir?" My computer upon being booted up and into existence is effectively saying: "At your service, Master, I'll do exactly as you say!"†

.

The only major problem with my interactions with my silicon slave has to do with my speaking clearly and in a language it can understand. Labrador Retrievers are the breed of choice for guide

---

* I must be in some spam database filed under *hermaphrodite*, for I get about an equal number of email offerings to expand my penis as my breasts.

† I did program my Amiga computer when I first got it and played with its voice emulation software to say something to this effect. What can I say?

155

dogs, and are considered one of the most intelligent of breeds, yet they rarely take 1st Prize in Obedience Trials. It has been suggested (mostly by fans of the breed) that the reason for both is that Labs insist on exercising independent judgement. If the blind man instructs his guide dog to take him across the intersection because he has heard the crossing signal's auditory beep indicating that the light is green, but a car is running the light, the good guide dog refuses to obey the command. A computerized robodog would just lead its master out in front of the speeding vehicle.

.

If you tell you computer to format your hard drive and erase all its memory, it's happy to comply. "You asked for it, Master." Your every wish is its command, even if you don't really want what you wished for—or you misspoke yourself.* In short, the trouble with computers is they do what you tell them to.

.

And exactly what have we told them to do that has directly affected—and will continue to affect—creative activity? A detailed answer to that question would take volumes, but one could say it all comes down to the *potential* for increased *productivity*. There are two different ways computers can do this: one is by allowing us to do things faster and more accurately; the other is by allowing us to do something that before we couldn't have done *at all*.

.

Doing Things Faster. Shakespeare's prodigious productivity is astonishing. He wrote 38 plays, 154 sonnets, and two long narrative poems in the short creative writing span of 22 years. Imagine what he could have done if he had a microcomputer with a word processor instead of a mere quill pen! Well of course it may be that it wouldn't have actually resulted in any more great plays or even any improvement to the versions we have.† An individual's creativity isn't a bottomless well where a bigger, faster pump will necessarily deliver up more to quench our aesthetic thirst. However, it is true that all creative endeavours involve some element of drudgery. If one can

---

* Of course the computer can be programmed with buffering questions to avoid such major disasters, but if you insist—well you can't blame it. The following parody of this failsafe system is never going to happen. User types: "Format C: drive". Computer prints to screen message: "This will erase all data. Are you sure?" User: "Yes" Computer: "Are you really sure?" User: "Yes" Computer: "Are you really, really sure?" User: "Yes" Computer: "Are you really, really, really sure?" User: "Yes" Computer: "I don't believe you. That is a stupid idea. Operation aborted."

† If his word processor had a spell-checker...

assign the drudge work to a 'slave', *and* one can use the freed up time to do more creative work, obviously there is going to be an increase in productivity.* Leo Tolstoy's long-suffering and devoted wife recopied by hand three complete drafts of *War and Piece*, while he simultaneously edited it. It is not unreasonable to believe that the time he saved by this spousal devotion (or abuse) made possible the completion of later masterpieces such as *Anna Karenina*.

.

To continue to focus on writing as an example (partially because I know more about this from personal experience), an important caveat to keep in mind is Postman's remark about being careful to use technology but not be used by it. Simply because one can type faster than write longhand doesn't mean you will produce more of worth. One can dictate even faster than type (and then have speech-recognition software turn your utterances into a text document on your personal computer), but I don't know of any significant writers that can—or do—work this way. In fact a surprising number of writers even still write longhand. That one can pump out more words in the same period of time can work against those words being carefully thought out, and so in the end might involve more time in editing and cutting and correcting—or, even worse, in the false confidence that no editing and revision are needed.

.

Related to this is another example of the pitfalls of relying on the computer to handle chores the writer considers drudgery: i.e., the reliance on one's word-processor's spelling-checker and grammar-checker. While this is usually a wonderful help in editing a manuscript, it can't be entirely trusted—yet something students seem to repeatedly do when preparing term papers. Although these great add-ons to word-processing software can quickly flag and fix 90% of one's straight-forward typos and grammatical goof-ups, one still has to reread and edit by hand. Putting excessive confidence in the computer's ability to take full responsibility for finding your mistakes, or even somehow miraculously improving your style, just does not work.

.

Nonetheless, overall the increase in productivity made possible by having a personal computer slave with access to the libraries of the world through the Internet can't be over-estimated. I will again beg

---

* Painters until fairly recently have traditionally assigned some of the tedious bits on large canvas to their apprentices. CEOs delegate responsibilities, rather than micromanage.

the reader's indulgence and fall back on personal experience. When I graduated grade school, I was given what I'd requested as a graduation gift: a portable manual typewriter.* I took an instructional book out of the library and spent the summer teaching myself to type—reaching sixty words a minute before I started high school.†
When I first began writing poems for possible publication, I would retype them ten or twenty times as they went through revision after revision—each time editing the 'hardcopy' with a pencil and then going back to the typewriter. This procedure became progressively more time consuming when I started writing fiction and other prose works. (An example of this obsessiveness is deciding it was necessary to change the name of a character in a twenty-page story and so retyping the whole damn thing.) When I got my first personal computer with word-processor and printer‡, my working world changed beyond recognition. Printing versions, the latest drafts, was keystroke away. And making changes disconcertingly (and sometimes dangerously) easy. And, as an example of how we can become personally, emotionally involved with our technology, I used to leave my electric typewriter on, humming reassuringly away in a corner of my study, while I worked at my computer.

.

When one's creative project involves research, the effect of computer technology is immeasurably greater. When I first started teaching, writing any psychology 'literature review' paper involved going to *The Psyc Abstracts* in the library. Allegedly every major refereed journal publishing scientific findings in the area of psychology is indexed by this resource. Every published paper has (usually right after the title) a

---

* The three possible career choices I was considering at age twelve were baseball player, scientist, and writer. I already had a nice baseball glove and a fancy chemistry set, so it was the only tool I was missing. It was a very necessary tool, since I had almost failed penmanship in school and hated writing longhand.

† In high school, being lazy, I took "Typing" as an elective, and was able to do all the course work in the first few weeks of the first semester, and of course 'earned' star status. Pride cometh before a fall. In the second semester a girl, who was destined to become a concert pianist and started the course never having touched a typewriter, beat me out in the final exam by I can't remember (or don't want to) how many words per minute. When she typed it sounded like she was playing a scherzo movement on a harpsichord. Maybe it was at this point I first glimpsed the importance of genetic gifts and transfer of training.

‡ I want to give credit where credit is due. This was a Commodore 64. The word-processing software was called *Paperclip*. The printer was dot-matrix without true 'descenders'.

one paragraph summary ('abstract') explaining what was done and found. These abstracts were collated, indexed by key words*, and published in huge volumes that one had to laboriously search through to find whatever one was interested in reading about. A comprehensive 'lit search' that took months in those days can now be done in an afternoon, for now *The Psyc Abstracts* are in a relational database called *PsycINFO*. The indexing is better and the abstracts can be downloaded onto one's computer for perusal. Researchers' lives have become one hell of a lot easier.

.

The hunting through my personal library and other libraries that would've been required to write this book (which flits and dances across so many domains) would probably have made this project too daunting for me—even for me at my most manic.† I simply can't imagine doing it with a typewriter and no Internet access, even though, of course, most books until very, very recently were written without these tools. I'm not alone in this: many other writers have admitted developing a similar dependence on—and appreciation of— their computers and the Internet.

.

Now, I haven't yet said anything about the effect of advances in computer technology in terms of scientific productivity. I don't think much really needs to be said here, for at least regarding efficiency, the effects are obvious. Remember, too, that a computer was originally a machine that just computed, merely replaced human 'computers', and did the tedious math. And number-crunching is a crucial part of much of science.

.

To again choose psychology as an example, the statistical analysis of data essential to the field has become exponentially easier. No psychology major can avoid learning to use the ubiquitous analysis software SPSS (Statistical Package for the Social Sciences). The first version was released back in 1968 and ran on university mainframe computers. In 1984 a version was released that ran on PCs by reading text files that were basically little program files. In 1992 a version was

---

* These keywords were often somewhat less than intuitive. If you were interested in the psychological effects of a child being taken from his mother at an early age, you had to search "maternal deprivation". *Psyc Abstracts* supplied a thesaurus to help find the appropriate key words.

† "So damn technology!" cry my more critical readers.

released that used the Windows interface, and the current version is as user friendly as such a behemoth can possibly be. Now most social science students are running a version of SPSS on their personal computers.

.

However, that one can do more and do it faster and easier has risks. Psychological research papers have become clogged with excessive and often confusing, even confused, statistical analyses. (This is another example of feeling that if you have the technology that *allows* you to do lots of things, you *must* do them. Who of us, if behind the wheel of a sports car that can go 200 km/hr, isn't tempted to ignore that silly 100 km/hr expressway speed limit—even if we don't really know how to handle a car at such extremely high speeds? Statisticians reviewing published papers, including those in the prestigious journal *Nature*, often find glaring errors. I am not at all expert in statistical analysis, but I do understand (and teach) the basic principles. Even I often spot egregious conclusions based on dubious statistics in refereed journals. These errors are somewhat understandable because the analysis software distances you from the underlying logic.

.

I see this all the time in my students' work. Consider one of the most basic of statistics: the mean. This type of average is calculated by adding up all the numbers in a set and dividing by how many numbers you have. We all learned to do this in grade school. If you got a 70%, an 80%, and a 90% on your three arithmetic quizzes, you averaged 80%. I've seen my students collect data in a lab experiment where the largest datum is 10 and then, after punching the numbers into their calculator or even SPSS, report the mean as, for example, 22! Often when I point out the absurdity of this result, they get indignant, saying their calculator or the computer *cannot* be wrong, and they are sure they entered the data correctly.* Well it's a lot harder to catch an absurdity in a MANOVA† analysis and a lot easier

---

* The way I usually approach teaching elementary statistics is to use simple numbers and make them do it all by hand so they get an understanding of the underlying logic of the statistic. This has the added benefit of their appreciating the computer when they get to use it, because it makes their life easier. If they'd started on the computer, I'd just hear complaints about how computers are too complicated.

† For those that care, this is an acronym for multivariate analysis of variance, used in studies with more than one change being measured; i.e., numerous dependent variables. This is tricky because if you measure enough things you're bound to find something "statistically significant", and there is the question of interactions between variables.

to just trust data entry and trust you correctly instructed SPSS—*and* that you are correctly interpreting the complex, detailed output.

Nonetheless, despite the risks, there is no question about the profound effect on scientific progress that the pure computational power available for scientists on their desktops (not to mention the new super-computers at their institutions) has had—and will continue to have. And this effect is not only in better pure number-crunching, as the earlier discussion of fractals makes obvious. The exponential improvement in the graphical capabilities of computers is equally important.

Doing New Things. Computer technology not only allows us to do things much faster than we've done before. It also allows us to do things we simply could not have ever even done before. (Sometimes the latter is because of the former, as with the generation of fractal images and many iterative depths.)

Two major components of all art are its creation and its distribution. Computers have revolutionized both. Computers have removed so many human physical limitations to the expression of creative ideas that vast areas of artistic endeavour have been opened to people who fifty years ago were denied access because of such fundamentally trivial things as poor coordination or not having years of specialized training and practice. Computers also have removed so many of the physical limitations of distribution and appreciation of creative work by the increases in reproductive technology.

The creation of art is the transformation of idea into object. Most people remain, justifiably, in awe of those who have the refined special sensori-motor skills to do this with apparent ease. We are all impressed with virtuosic piano performances, but those who have taken piano lessons are much more impressed, for they better realize the difficulty, the hours of practice behind it—and the native gift that made such virtuosity possible. Similarly with visual art, admiration of precision of technique, of draughtsmanship, is almost universal. Again, for someone who has taken a drawing class, and whose attempts at perspective drawings look like a catastrophic distortion in space-time, the works of the Renaissance masters justifiably inspire awe.

We must, however, not forget that what *ultimately matters* is the art work, not the creator of it, and certainly not his or her exceptional and specialized skill in transforming idea into art object. I've projected various high-resolution realistic landscapes and portraits on a large screen for my students in my Psyc of Art class and asked them to tell me whether they think they are photographs or paintings. They usually don't guess better than chance. What is interesting is that if I tell them an image is a painting, they are very impressed with its 'artistry', but if I tell them it is a photograph they almost shrug. And sometimes I lie. Of course there is no reason we should think more highly of an artistic image just because we think it required the sensori-motor skills of the painter.

The creative idea matters. Cameras made possible the earliest transformation of visual idea into visual art object available to those who were not good draughtsmen, but it is the computer that has really opened up the creation of visual art to those lacking even the most primitive drawing skills. Many of the early innovations in Modern Art, such as collage and found art, were the first step in this direction, but the introduction of personal computers with powerful graphics software is what has totally changed the visual art world. I speak from personal experience, for I have neither natural gifts nor any training in the usual drawing and painting skills a visual artist 'should' have. But I have learned to use sophisticated graphical software, digital cameras, and also learned enough programming skills that I can produce complex art prints that at least some galleries value enough to exhibit and some people even buy. And a friend of mine who no one who isn't tone deaf would want to listen to try to play the guitar is using his computer to create quite impressive musical compositions.

Computer technology has empowered people with the tools to transform their creative ideas into art—without necessarily having inherited and then laboriously nurtured the traditional skills to do so. Still one has to remember that whether the tools of transformation from idea to art are the result of inherited ability and a long painful period of apprenticeship or instead the result of learning and applying the new technology, the creativity precedes this. There are many painters whose draughtsmanship is better than even that of Rembrandt or Picasso or Dali, but whose work is insipid. There are many people who have mastered *Photoshop* and computer image manipulation techniques who are totally lacking in any creative ideas.

The second huge effect of computer technology is in the area of distribution of creative works. It is one thing to create something and quite another to find an audience for it. The problem of distribution is very different for the different arts.

I think that for literature there are four really major 'revolutions' or paradigm shifts in its distribution, all a result of advances in technology.

- Writing itself
- The printing press
- Paperbacks
- Electronic storage

Before writing, of course, all 'literature' was oral and distribution was by word of mouth. Given the fallibility of memory this meant all literature was undergoing constant revision. Given the limits of transportation, all literature was local. Both these obvious limitations are exemplified by the great myths. For example, myriad variations of the Great Flood Myth occur in the majority of past cultures we know anything about. The invention of writing stopped the revision and freed distribution from being linked to individuals and individual memory.

Like writing, the invention of moveable type and printing so obviously important, it hardly need be remarked. Writing stabilized literature; printing allowed the stable work to be replicated and more widely distributed.

What I call the 'paperback revolution' I believe is a less obvious and often under-appreciated event. Its effect was to greatly expand the distribution of literature across economic borders and barriers, for it was another great leap in reducing the cost of owning literature. The production of cheap copies of bound literature on inexpensive pulp quality paper can probably be said to have begun and gained momentum around the middle of the nineteenth century with the "Penny Dreadfuls", stories published on cheap pulp paper and sold for a penny. Many of these were indeed dreadful literature, lurid stories aimed at working class boys—but not all of them. It was about a century later that the modern pocket-sized paperback book started being mass-marketed in North America.* While it is true that a mass market implies appealing to mass taste, it is also true that, given low production costs, a really massive market includes enough consumer diversity that it is still profitable to aim well above the lowest common denominator. From the middle of the 20th Century on, one didn't have to be a wealthy aristocrat to have one's own specialized library. When I was in high school in the sixties, we had regular paperback book sales in the school auditorium, with the books being chosen for having some literary value.† It was a revelation to me that I could own my own library where I could go back to a book whenever I wanted, rather than back to the public library.‡

.

Accessibility is related to cost: as cost goes down, accessibility goes up. Before Gutenberg the cost of a single Bible laboriously copied out by some monk has been estimated at equivalent to 500 thousand dollars. (And no doubt each edition had different 'typos'.) By the time I was growing up one could buy a book of the same length for 50 cents—and Bibles were being given away. Now most significant books produced more than fifty years ago are available at virtually no cost at all, because of the fourth of the revolutions in literary distribution: digital storage and networked distribution of text. This is a subject to be explored in more detail in a later section, but suffice it to say here that it is now possible to do something never before possible: the writer can make his or her work available worldwide at very little cost. This is a totally new situation of great significance.

.

I've only considered the literary arts, but this new status for distribution of art applies to most of the arts. There are important differences from domain to domain, but the general effects of this expansion of the marketplace apply across the board.

---

* Often credited as the first real 'paperback', Pearl Buck's *The Good Earth* was released in this format in 1938.

† In those days school authorities were more concerned about exposing us to ideas than shielding us from ideas that might offend parents. I remember that one of my early purchases (for 50 cents) was a copy of Bertrand Russell's *Why I Am Not A Christian*.

‡ This convenience eventually became inconvenient as my personal library grew and grew and grew. Now, although I have a reasonable size house lined with bookshelves, I still have to rent external storage space to house some of my library. And of course the books boxed away in a warehouse are not exactly what one would call easily accessible.

In the visual arts, it is possible for artists to put on display to the world reasonable reproductions of their works created *in any media* by digitizing the image and putting it on the Internet. No longer is the audience for new work limited to those who can go to a particular gallery at a particular time. The increase in accessibility resulting from computers and the Internet is even greater than in literature: new literary work published by a small press wasn't easily accessible, but it was still more accessible than new visual art.

Increased accessibility to music independently and inexpensively produced is another variation on this story, with interesting twists and turns. I won't venture into that heated discussion. I'll just hope I've made my point that computer technology has introduced something totally new in terms of accessibility and distribution of creative products. Its complicated effects will be explored well into the foreseeable future.

Turning now to the effects on science, the first thing that should be noted is that the just discussed increase in accessibility and ease of distribution applies here as well. I'll postpone a closer examination of this to a later section on the Internet.

Simulation! That is what I speculate to be the most significant new territory opened up by computer technology. This is not surprising since simulations are mathematical models, and math is the heart and soul of a computer.

'Simulation' originally had, and still does in some ordinary usage, a somewhat pejorative connotation, for to simulate is to 'fake' or 'feign' something. A simulation is not the *real* thing. A simulation is a model. In science, it is a scientific model—a concept that is often misunderstood. Models, of all kinds, are similar, simulate, some features of what they model, but not all its features. If original and model were alike in all features, the model would of course be the thing itself. You can buy a kit to build a very detailed model airplane, but when the glue has dried you can't climb in it and go flying about the skies. Scientific models are like that as well, yet they are sometimes irrationally denigrated for not being identical to the thing modeled. For example, cognitive scientists design computer models of human cognition, but oh so often I've heard a student say in objection to such models "Hey, my brain isn't a just a computer!" Of course no cognitive scientist writing a program to simulate human cognition thinks our brains are a bunch of transistors or even that computers 'think' the same way we think, any more than a meteorologist writing software to model weather patterns thinks his computer is full of air and clouds.

.

What simulations, computer models, allow one to do is try out something empirically that would in reality be impossible or impractical—or highly undesirable. One of the early major computational simulations was the Manhattan Project's mathematicians' and physicists' modeling of the first nuclear bomb. Remember that science is just a game of 'let's see what happens'. Computers make playing this game a lot safer. Of course, the simulation program might be flawed; and if the game can—and is—played out in the real world such 'bugs' will be very evident and may be disastrous. Usually, however, the stakes are not as high as when predicting whether the first nuclear detonation will 'only' kill all living things in a five mile radius or whether it will instead wipe out all life on the planet.

.

A more innocuous—and more typical—example is testing a probability theory. The game Yahtzee involves rolling five dice to make certain combinations that score different points. Like all dice games, knowing the probabilities of certain combinations gives the player a great advantage. One evening I was fooling around with applying simple probability principles to predicting the likelihood of certain five dice combinations. Unsure of my logic for one set of predictions, I programmed my computer to simulate rolling five dice *ad infinitum* and keep track of the percentage of times the particular combination occurred, printing this percentage to the computer monitor. I ran the program and watched the percentages scroll by. At first my theoretical prediction and the empirical output were, as I expected, quite different, for it takes time for matches between theoretical probability and empirical results to come together.* Soon I saw my calculated prediction and the actual result come quite close together, and I felt vindicated in my reasoning. I left the program running and went to bed. In the morning when I got up and looked at the output scrolling by, I was a bit disturbed. The output I'd theoretically predicted was only off from what I was getting from my computer simulation at the fourth or fifth decimal place.

---

* No one really expects that rolling a fair die six times will result in each side coming up only once, but if after sixty thousand rolls, you don't have very close to ten thousand each of every number from one to six, something is fishy.

Nevertheless, given the huge number of rolls that my indefatigable computer had tossed overnight, this was still way too large a discrepancy. I let the program run all day, checking it more and more frequently, and still the simulation output refused to converge on my theoretical prediction. So I went back to my calculations and, needless-to-say, found a small error. I fixed it and ran the simulation again. This time the theoretical prediction and the empirical simulation output came into agreement at seven decimal places in a very short time.

.

There is no way I could have found time to 'roll dice' anywhere near as many times as my computer did, and had I actually tossed dice for a mere few days, I would have wrongly concluded, when seeing the initial approximate convergence between predicted and actual outcome, that there were no flaws in my logic.

.

Computer simulations now are common in virtually every field of science. It is interesting that they are especially useful in areas where eventual real-world empirical testing of their predictions is totally impossible. Two good examples are cosmology and evolution. This may at first seem strange, since one will never be able to manipulate variables in the real world to test if the model is accurate. However, what one can do is see if the simulation predictions for known past events matches the real world data. If it does, then inductive reasoning permits one to boldly assume it will hold for the untestable future predictions.

.

Computer simulations are so powerful and fascinating that they have filtered down to popular entertainment software. For example, *SimCity*, which allows one to design and build a city and test its viability against natural disasters was an extremely popular software program ever since its introduction back in the days of C64 personal computers. Really, in one sense, any computer game where strategy is as important as reflex speed is a simulation game. Like in science, one tries different strategies to see which works—or works best.

.

As computing power continues to increase, what were once thought-experiments can be transformed into virtual experiments conducted in the virtual laboratory that computers have become. So now some skill at programming (i.e., clearly instructing one's computer slave) is approaching the importance of mathematical skill in the creative and addictive game of science. If you give someone an interesting program, you'll keep them occupied for weeks, but if you teach them to program you may very well keep them occupied for a lifetime.

THAT SEXY SYNTAX ERROR: PROGRAMMING AS ART

Stereotypes can create wonderful images. Consider the artist with his bottle of wine and studio littered with discarded sketches. Similarly, visualize the computer programmer with his bottle of coke and workspace littered with empty pizza boxes and cryptic printouts of program drafts. And, like most stereotypes, these romantic images have some grounding in reality, albeit often a past reality.

.

However, a strange transformation has been occurring: it seems there finally is something sexy about being a nerd. Perhaps because computers and other technological gadgets and gizmos have become so much a part of our lives, those who actually seem to understand how they work, how to make them work, are seen as gurus. And gurus always have been sexy. Someone* remarked that "managing programmers is like trying to herd cats". Well, individualism is also sexy.

.

Forget pocket protectors and thick glasses. There is a new stereotype: Silicon Valley is full of rich, hip, fit folk driving around in expensive sport cars or pedalling around on high-end bicycles. It is a mystery how they manage this despite supposedly doing all-nighters living on Pepsi and pizza, but it certainly adds to their charismatic image and desirability.

.

Most programmers take what they are doing very seriously and are every bit as passionate about their calling as are artists. Many strange things get honoured with the label 'art' by their practitioners. The usual justification they give is that what they do requires a special skill that cannot be reduced to some mechanical instruction set (what a programmer would call an algorithm†) and requires intuition and insight. However, the working definition of art I've adopted demands more than this: Where is to be found, it is reasonable to ask, the aesthetic experience in a computer program?

.

There are two answers to this question.

---

* Allegedly Greg Settle.

† Of course what programmers do is create algorithms, but the reduction of some complex thing to a simple mechanical formula is the creative part, the art: an engineer applying Boyle's Law is not being creative; Boyle was.

The first involves some minimal grasp of the nature of programming and its relationship to mathematics—as well as some understanding of how a mathematical proof can cause an aesthetic response in someone. Writing an elegant computer program is like writing an elegant mathematical proof. In fact, both math and programming share the concept of elegant algorithm: i.e., a simple, fool-proof instruction set that always leads to a solution. Of course, aesthetic appreciation of this is what I've previously labelled 'deep', and is only accessible to those 'in the know'.*

The second answer is that a computer program can produce an aesthetic response even in those who haven't the vaguest idea of what is involved in creating the program. An obvious example is a program that is specifically designed to be an art 'object'. Maybe because of the relatively low status of art in most highly technological nations such programs are peripheral to our lives—much like the Mozart concerto played softly as background music in elevators.

An example is the so-called "screen-saver". These programs were developed when computer monitors could get permanently 'tattooed' by an image that didn't change over long periods of time. These programs solved this problem by cycling through images that would eventually randomly spread out the pixels being displayed.

Screen-savers are a truly innovative development that is under-appreciated. They present new images every few seconds, and this attracts our jaded attention. Most of us no longer really notice the paintings on the walls of our homes—works that we purchased because they once so moved us we were willing to open our wallets. But were some household slave to come in after we'd gone to sleep and replace each of them with other works we'd bought (but had to leave in storage for lack of wall space), in the morning we'd again look at the art on our walls with real appreciation. And when these were, in turn, replaced by the previous artworks, those works would seem fresh.

This phenomenon is called 'adaptation' or 'habituation' by psychologists. Our senses adapt and habituate to stimuli that are ever present, so we no longer notice them. We only again pay attention

---

* It is said that the reason 'real' programmers don't comment their code is because if it was hard to write, it bloody well should be hard to read.

when things change. I once, not that long ago, speculated about the day when the artwork on our walls would be regularly changed without us having to do anything, because it would be artwork displayed on a flat screen monitor hanging on the wall, and the household slave responsible for changing it would be a computer. Well of course that day has come, and the cost of a programmable flat-screen digital art display unit is now equivalent to an admission ticket to an Art Museum.

Then there are programs with a real, practical function that because of their design, appearance, and smooth functionality evoke in the user what can only be called an aesthetic experience. By analogy, consider the car connoisseur. For example, I certainly don't have difficulty in accepting the claim that a perfectly preserved '57 Chevy is a work of art. Admiring and driving this classic car is indisputably an aesthetic experience. And as previously argued, practical functionality does not exclude something from the category 'art'—in fact, it can be a part of what defines it as art.

So, to return to computer programs, compare two once widely-used email programs—the one called *Thunderbird* and the one marketed by Novell called *Groupwise,* which some institutions (including the one I teach at) used. *Thunderbird* may not be high art, but *Groupwise* was an aesthetic atrocity—partially because it looked ugly and partially because it was clumsy to use. Both would do the job, but we've learned to expect more than bare functionality. You can drink you vintage Cabernet Sauvignon from a beautiful crystal wine glass or from a Styrofoam cup. In computer terminology one might call the vessel used to deliver the goods an interface.

"Powerful program, but what an ugly and clumsy interface!" the computer user often exclaims. As in the design of any thing with a function, we value form complementing function. We value form even if it only *decorates* functionality. A chair that is beautiful to look at but uncomfortable to sit in is unsuccessful functional art, but still may be art. A chair that is comfortable to sit in but looks like it was dredged up from the local landfill site is just unsuccessful art.

Computer programs can be art, but they are flawed if they are not functional, *and* they are also flawed if they are ugly. There is a whole field of research that deals with this: it is called HCI, an acronym for Human Computer Interface.

Of course there will always be those who claim to just care about functionality, with no interest in 'superficial' aesthetic issues, and it *is* true that functionality in most cases should have priority. But even these folk will prefer the nice looking chair to the ugly one if both are equally comfortable.*

The design of Graphic User Interfaces (GUIs) are largely responsible for the growth in popularity of personal computers and interest in the Internet. The cynic might label it as nothing more than fancy packaging, but nonetheless it is grassroots interest that has fuelled the development of computer technology. The home computer only became really popular outside of nerd culture when GUIs made using it far easier than typing in cryptic commands after a blinking cursor. The Internet, and especially its GUI incarnation known as The Web, has been brought to us by grassroots' interest in seeing images of naked women.†

What I suspect will move to centre stage in the future, because it is already grabbing a lot of attention at the grassroots level especially among young people, is the development of virtual reality and interactive computer interfaces. This is a programming art form that has already gained respectability (of sorts) in the games and entertainment sector.

Many artists now are studying programming, just as artists in the past studied geometry and perspective, to learn to use these new tools to create entirely new art forms. Exhibitions of *avant-garde* art more and more involve the use of computers and modern technology. For many contemporary artists being able to program (in the broad sense of knowing how to interact with the new computer technology) is

---

* A parallel situation is in choosing one's doctor. One can claim to not care about bed-side manner and only want the most competent physician, but if the choice is between two equally competent doctors, who wouldn't choose the empathetic over the arrogant and abrupt?

† Those of us that surf to the great art museums of the world and sites with the latest news on scientific discoveries really should pay homage to what has made all this possible. The original Internet, the Arpanet, was developed and funded by the military establishment to support research that would help create more weapons of mass destruction. The latest graphical incarnation of the Internet we call The Web owes its existence to the popularity of pornography sites. Anyone doubting this can check any site that lists the most accessed sites. More than 90% are 'sex sites'. Interest in death and destruction and sex has made possible listening to digitized Beethoven or viewing digitized Rembrandts on our home computers.

becoming as important as being able to draw in perspective was for the Renaissance artist.

WHAT A TANGLED WORLD WIDE WEB WE WEAVE

When I confessed to a friend how much time I spent entangled in The Internet, he laughed and said how lucky my wife was not to have me bugging her all the time. Then he started riffing on the theme. "Give a man a fish and you feed him for a day; give a man Internet access and he won't even be interested in eating for a week." "The way to a man's heart isn't his stomach anymore, it's the information highway." "Does your wife call you for dinner by sending you an email?" He was being a bit unfair, since my wife probably spends more time online than I do. However, it is a fact that my wife and I really often do communicate by email, even though her study is the room next to mine, and we can easily talk to each other without rising from our desk chairs. It's just easier to forward an interesting article or some document one of us is working on than to try to describe or read aloud the piece—or print it out and carry it over the adjoining room. Also it is less disruptive, for one can deal with it when one is at a natural break in what one is doing. Email is much better than phone or direct oral communication in this way.

.

How on earth did all this happen? Who would have predicted, even a few decades ago, such a wide-spread and bizarre change in modes of communication? Even when microcomputers began to appear in people's homes, what was touted as their practical uses for the average person were things like keeping a database of one's favourite recipes. Of course this wasn't really practical as few people would laboriously type recipes into a database program on their Commodore 64.* Now, however, virtually everyone with a personal computer does have a huge database of recipes at one's fingertips— on the Internet. I'm sure our family isn't atypical in checking this database before deciding how to cook up that trout in the fridge.†

.

The introduction of the personal computer was a huge revolution, but its potential and power outside of specific domains only became apparent as the Internet developed. This topic is so broad and has

---

* The oft-quoted but mysterious personage named Doug Larson allegedly quipped that "home computers are being called upon to perform many new functions, including the consumption of homework formerly eaten by the dog."
† A less trivial example of this is the new project initiated by E.O. Wilson to build "The Encyclopedia of Life Project", intended as an online reference source and database for the world's 1.8 million named and known species and to facilitate the discovery of those yet unknown.

been the subject of so much in-depth analysis, I'm only going to touch on two implications that seem particularly pertinent to the topic of human creativity: new possibilities for creation and new possibilities for distribution.

.

Once again it was science that was the agent behind the scenes that set the stage for the Internet revolution. When the USSR launched Sputnik in 1957, the collective self-esteem of the people of the United States took a beating, and this apparent example of Soviet technological superiority combined with Cold War paranoia led to the Americans re-emphasizing and supporting science education and scientific research. The following year the Advanced Research Projects Agency (ARPA) was created. The importance of communication in doing science is greater than in the arts. Scientists need to know what their colleagues are doing, discovering, for scientific knowledge is itself a huge interdependent network. So in 1983 The National Science Foundation in collaboration with ARPA constructed a university network backbone and a few years later this network expanded so that access became available to commercial interests and the larger scientific community. It then grew like Topsy, metamorphosed from ARPANET to the Internet, but although it greatly augmented communication among scientists, it remained difficult to use for the uninitiated. But in the early nineties the World Wide Web came into existence. Here is where the agent of art stepped in: graphics again (HCI) were the catalyst to trigger exponential growth. The Internet then branched out as wildly as synaptic connections do in the developing human brain. And the number of people climbing about this tangled web also exploded. All current estimates of the number of people using the Internet exceed one billion!

.

Since it was scientists who first realized the creative potential of the Internet, I'll deal with them first. Initially the way this new communication tool increased creative productivity in the scientific community was through nothing more than expedited communication. One didn't have to wait for publication in refereed journals to find out what one's colleagues were up to, for now researchers geographically isolated could easily and effectively keep in touch as they proceeded with their research. Even in regard to published findings, the Internet has made finding material phenomenally faster.

.

Another way scientists, especially social scientists, have used the Internet is to collect data. It is a lot easier to set up a website questionnaire which interested parties can respond to than to recruit students at a university or shoppers at a mall to answer a bunch of questions—and sometimes somewhat personal questions. Half of the data I collected for my study on aesthetic and mystical experiences came from Internet users who filled in an online questionnaire. Of course there are methodological sampling problems with this, but so are there with conventional paper and pen methods.

.

This is now, and while no doubt this increased ability to communicate and collect data will continue to grow, what I see as the major future development, what I envision as really new scientific creative territory is the use of networking individual computer power. The microcomputer on your desk is probably more powerful than the hug mainframe computers at major research and educational institutions only a few decades ago. The newest mainframe super-computers, such as the famous Cray computers, are almost incomprehensibly powerful, but I would guess that the ratio of their power to that of the average personal computer sitting on the reader's desk isn't that different than the ratio of the power of a seventies mainframe to that of first Commodore 64s. Size matters, but numbers matter more. A bee is no match for a human in one-on-one combat, but a swarm of bees can easily kill a human. If the computer on my desk is only one thousandth as powerful as a modern mainframe computer, should my computer join forces with two thousand other computers, our combined power is twice that of the big guy in frugally doling out computing time to academics at some large university.

.

One example of the way this principle was applied was the *PiHex Project*. By distributing computing power among approximately 2,000 microcomputers which ran calculations in the background while not engaged in any other task, this distributed computing project was able to calculate the deepest calculation of the value of pi to date—at its termination in 2000. Currently, there is a similar project to find Mersenne prime numbers.* One downloads the program from a website and runs it as a background process. There are cash prizes for lucking out and finding new ones!

.

---

* A Mersenne number is a number that is one less than a power of two. A Mersenne prime is a Mersenne number that is also a prime number.

In art, the already mentioned access to images and text and auditory material is continuing at a phenomenal pace. For an artist to have ready access to all this wonderful stuff from which to create new work is a truly amazing resource, stimulus, and catalyst.

.

Then there is the increased ease of distribution of created work made possible by both personal websites* and such sites as *YouTube†*, where anyone's video art can be uploaded for viewing by anyone wired into the Internet. Similar sites exist for new musical creations.

.

However, what I would even suggest there is an expansion of creative territory. A website can be an *objet d'art* in its own right—and often an interactive art object. This is only beginning to happen, but I'll predict that it will become more and more common. As the original HTML programming protocol for websites evolved, the design of them became a new art form. They weren't just a place to see or download art: they were the artwork itself. The real world Guggenheim Museums in New York and Bilbao are both artworks themselves and not just a place to exhibit art. It was inevitable that eventually websites would be created as stand-alone works of art with no real-world referent.

.

With high speed access becoming more and more common and the development of highly sophisticated tools to create intricate, animated and even interactive websites‡, the Internet site has emerged as a new type of art object. Like any new art medium, it has suffered—and continues to suffer—some growing pains. The web artist has had to deal with unique problems.

.

For one thing, it is impossible for the artist to be sure that what he sees as he views his web site creation is what every user will see—in fact, it is absolutely certain that they all won't see what he sees. Different computer monitors display at different resolutions, have different capabilities in terms of the number of colours that can be displayed and are not identically colour calibrated. Different browsers

---

* My own digital art can be viewed at my Internet domain: KenStange.com

† Its importance was acknowledged when *Time Magazine* in 2006 named *YouTube* "The Person of the Year".

‡ An example every web designer knows is *Dreamweaver* with its powerful add-ons for programming complex graphics and animation.

also have differences in the way they interpret that code that produces what the viewer experiences. Recently things have become more standardized, but the way different browsers interpret the code they receive still differ enough that the good website designer has to test his site on the major ones. The differences in users' resolution and colour capabilities has also decreased, but remain far from completely standardized, and what is the norm now is likely to be replaced as more and more people upgrade their computers.

.

Another issue is that of permanence. Although digital information does not degrade, the website as art object only exists virtually. If an artist paints and sells a picture and then dies, the picture is still out there. If an artist creates a beautiful website and dies, he'd better hope his estate keeps paying for the web-hosting to keep it from evaporating into the outer reaches of cyberspace. As anyone who tries to visit old 'bookmarks' for favourite Internet sites knows, the life expectancy of most websites is less than for small literary magazine. Unlike old lit mags (which still can be still be found in used book stores), most defunct websites disappear without a trace.

.

Furthermore, what if the Internet technology changes drastically? This is really inevitable. Will the programming and graphic code that gives the artist's Internet artwork substance become obsolete and not work—or not be used—any more? I experimented with computer art works on both the C64 and the Amiga. I still have the disks with these programs on them, but now my only audience is those who have these now antique computers that can read the disks and run the programs. During the early years of personal computers, programmers spent—or perhaps wasted—a lot of time 'porting' programs from one operating system 'platform' to another in a desperate attempt to keep up with advances and changes in technology and hardware.*

.

The kind of creative artist who is attracted to the Internet web site as new medium of expression must have some very special characteristics—characteristics more often associated with the scientist. Even with the powerful new design tools, creating a website artwork involves programming skills, doing math, and obsessive

---

* To some extent this issue of permanence has always been critical and intertwined with the current technology. The pointillists Seurat and Signac bought their oils from different companies, and Signac's paintings remain far more vibrant and truer to the way the originals looked than do Seurat's.

concern with detail. It involves testing and revising and fine-tuning. It is more like designing an elegant experiment than painting a picture. I wouldn't go so far as to say it is more a science than an art, but it has more science in it than most art.

.

Furthermore it requires a mindset that admits that whatever is created is in some ways tentative, just as are all scientific theories. The website artwork is too deeply rooted and based on current technology and too 'virtual' to have the feeling of solidity and permanence that art objects such as paintings or sculptures have. And it isn't physical even in the creation stage. Sitting at a desk typing on a keyboard in front of a computer monitor is very different from working in a studio with paint and canvas, or chisel and marble.

.

The one thing that can be said with confidence is that the evolution of the Internet website into art object is substantial evidence that the long overdue re-convergence of science and art is happening.

CASE STUDIES: MARSHALL MCLUHAN, STEVE JOBS

Even though most people would not characterize Marshall McLuhan or Steve Jobs as either artists or scientists, I selected these two individuals to represent the changing characteristics of creativity in our technological age because they both epitomize a new kind of creativity and have affected it. Marshall McLuhan was named as the "patron saint" of *Wired Magazine*—the premier hip 'technology periodical' which should be perused by anyone who doubts that developments in science and technology aren't of increasing interest to artist and intellectuals. He was so honoured not because of his insights into the computer revolution, for he died before it got into full swing, but rather because he was the first major intellectual to devote his life to trying to understand the effects, including aesthetic as well as social effects, technology was having. His approach was anything but academic (or scientific), and was very much in the spirit of the age he was trying to understand. Steve Jobs, as the founder of Apple Computers, of course, is one of the two major players in the personal computer revolution. While his domain of creativity is neither science nor art, his vision of the nature of the computer as a creative tool is a glimpse into the future.

.

.

Marshall McLuhan was born in Edmonton, Alberta in 1911. His father ran an estate business and his mother was a Baptist schoolteacher. He earned his BA in Arts and Sciences with distinction from the University of Manitoba, followed by an MA in English from the same institution. He then attended Cambridge where he studied under I. A. Richards and F. R. Leavis and was influenced by what their school of literary criticism, which was called "The New Criticism". It emphasized the actual text and insisted on ignoring any contextual and biographical information when analyzing a literary work. While he claimed this greatly influenced his subsequent thinking, most people would say that his own contribution to critical analysis involved quite the opposite approach: context seems of primary importance in his own analyses of the effects of technology.

.

He converted to Roman Catholicism in 1937 and by all accounts was quite devout for the rest of his life, although he kept his religious beliefs entirely separate from his creative and scholarly work. For the rest of his academic career he taught at various Roman Catholic colleges, eventually ending up at the University of Toronto's St. Michael's College. In the early 1950s he began the "Communication and Culture" seminars that drew wide attention. He published his first major work, *The Mechanical Bride*, at this time. It was an analysis of the effects of advertising and drew widespread attention.

From then on he assumed the mantle of guru of the new technology of the time. This very conservative, religious academic was embraced by the counter-culture which was fascinated by and in love with the new technology. The irony is that he actually viewed the effects of the new technology as more detrimental and dangerous to civilization than potentially enriching. Even the term that has been adopted as a symbolic of a new age of communication and cooperation, "The Global Village" had sometwhat pejorative connotations when McLuhan used it. He seemed to think it was more likely to lead to totalitarianism than some wonderful world-wide love-in—which is how the sixties generation misinterpreted his message, and so to this day the phrase continues to have positive connotations McLuhan never intended. The fact is that McLuhan feared that this turning of the world into a 'global village' could destroy individualism and lead to the ultimate form of totalitarianism: rule by the masses. The following quotation from *The Gutenberg Galaxy*, published in 1962, isn't what one would expect from someone who is welcoming the new information age.

"Instead of tending towards a vast Alexandrian library the world has become a computer, an electronic brain, exactly as an infantile piece of science fiction. And as our senses have gone outside us, Big Brother goes inside. So, unless aware of this dynamic, we shall at once move into a phase of panic terrors, exactly befitting a small world of tribal drums, total interdependence, and superimposed co-existence."

However, once McLuhan became established as the Oracle at The New Delphi of Technology, he softened his warning messages and became less dour and more playful in his prognostications. He published *Understanding Media* in 1964, *The Medium is the Massage** in 1967 and *War and Peace in The Global Village* in 1968. His analyses of popular technological media, which he had come to stake out as his

---

* "Massage" was initially a typo. It was supposed to be "message", but McLuhan had a sense of the power of the pun and insisted the typo not be corrected. "Massage" suggested all kinds of interesting word play, including 'mass age'. He subsequently often gave talks about its potential meaning.

own personal intellectual territory, became less polemic and more playful—to the point of seeming self-contradictory. Some serious academics appalled by—and envious of—his popularity tried to pull him down from his oracular pedestal—but with little success. The institute he initiated for the study of media and culture at the University of Toronto survives today. He had become the most unlikely folk hero of what was then the new technology. Here was a man who didn't seem to even really like the new-fangled technology. He was a devout catholic, a stodgy academic in an ill-fitting suit who mumbled when he lectured, a man who didn't even much care for watching television except as 'research' and just viewed all the monumental changes occurring around him as nothing more than material to do intellectual riffs on to the amusement of fans and the annoyance of those who took these changes more seriously—but who weren't themselves taken seriously in their less free-wheeling analyses of them. Paradoxically, the group-think and mindless collectivism he associated with the new media, as opposed to what he saw as the last bastion of individualism, the print media, was interpreted by many of the radical youth as a positive change, as support for a communal life-style and rejection of the boring old media such as books.

Whatever one may think of Marshall McLuhan's ideas (which deserve far more extensive and fair consideration than I've given them here), he could justifiably be called the father of the new field of critical thinking about new media, about new modes of communication and creative expression made possible by technology. As the magazine who honours him as their "patron saint" demonstrates, this attention to the implications of the new technology is itself a worthwhile creative endeavour.

After several years of ill health, Marshall McLuhan died in his sleep on New Year's Eve of 1980. This was four years after Steve Jobs had founded Apple Computer Company and four years before the Apple McIntosh Computer forever changed the technological landscape McLuhan had been mapping.

Steven Paul Jobs was born on February 24, 1955 to Joanne Carole Schieble who was unmarried and immediately put him for adoption. He was adopted by Paul and Clara (née Hagopian) Jobs whom he always considered his 'real' parents—distancing himself from his biological parents. Evidence of his early interest in computers was his voluntarily attending lectures, while still in High School, at the

Hewlett-Packard Company. After graduation he started attending Reed College, but dropped out after one semester. Again, typical of creative individuals, his rejection of formal schooling did not mean rejection of schooling or learning, for he still audited courses after dropping out. One of the courses he chose to audit also presages his interest in aesthetics: he sat in on a class on calligraphy, something which he credits with inspiring his insistence on different typefaces and proportional fonts on the Mac computer.

.

In 1974 he and his future partner in the original Apple Company, Steve Wozniak, joined the Homebrew Computer Club, and allegedly the two of them joined forces selling illegal "blue boxes" which emitted an audio beep at a frequency that permitted free long distance calls. At this time Jobs was working as a technician for the video game company Atari.

.

He then took a 'spiritual sabbatical' and backpacked around India, returning with shaved head and funky clothes, if not spiritual enlightenment. He went back to work at Atari, and continued to hang out with his brilliant, nerdy friend Wozniak, who had designed a personal computer for his own use. So at the age of 21, on April 1st, 1976. Jobs convinced Wozniak to partner with him in forming Apple Computer Company. The rest in entrepreneurial history. Four years later Apple Computer became a publicly traded corporation and Steve Jobs was a multi-millionaire.

.

Jobs is exemplary of the future of the personal computers and the Internet, for one major reason: he realized the importance of aesthetics. His counterpart, Bill Gates, may have shown more ruthless business acumen in the past, but he seemed to lack any sense of this whole new field as being intricately connected to art and design. From the very beginning of Apple and Microsoft, this difference in emphasis was evident. It is a gross oversimplification of both Gates' and Jobs' personalities, but I can't help but imagine if the two of them got off an elevator at some university and to the right was the science department and to the left the art department, Gates (with his pocket protector) would turn right and Jobs (with his sandals and chinos) would turn left

.

Steve Jobs was not a computer whiz. He probably did not score a perfect on his SATs as Bill Gates supposedly did. But he had profound aesthetic sense. (He also, like Gates and unlike most creative people, had a lot of practical business sense.)

Apple operating systems and hardware are noted as much for their elegant design features as Microsoft products are known for being absurdly complicated and clumsy. Microsoft programmers have tried, God knows, to erase this image and make things "user friendly", but they have certainly not been completely successful. For years I used a PC running the Windows-XP operating system, but whenever I'd install something that is very optimistically called "Plug and Play", I knew it was more a case of "Plug and Pray!" My prayers were only answered a bit more than half the time.

This is why I eventually switched to an Apple computer. I think the evolution of my own personal attitude toward and relationship to computers is typical of many people interested in the creative potential of computers. Apple is rapidly becoming the computer of choice; albeit for those that can afford it and the expenses and complications of switching from a Microsoft-based system.

Apple computers and their operating system were initially very 'closed architecture'. Everything with Apple has always been made to be very pretty and elegant, but if something went wrong, there was no way you could hack into the system to diagnose the problem and try to fix it. In one sense they were ultimately digital, ultimately binary, for it was a case of something works or it doesn't. Live with it. To a great extent this is still true: Apple products remain largely tightly sealed black boxes (although their products are often white). But less so than in the past. By contrast Microsoft's inner workings always have been open to fiddling with—although, alas, often to no avail. So for the nerd, the technologically inclined, the person who wants to modify things to his own personal specs, Apple products seemed boring and recalcitrant. They look pretty and were easy to use, which makes them appealing if you want them as an efficient, reliable tool. I used to call Apple's Macs "Yuppie Computers". They were so pretty and cute and easy to use—as long as the only uses you might have for them were the ones that were built in—and allowed.

Two conflicting forces are at play here, and both are related to aesthetics and creativity. For the person wanting a computer that is elegant, reliable, as well as highly functional when it comes to performing specific creative work, Jobs knew how to design and effectively promote such a product. For the person wanting more power to personalize or modify their computer as creative tool, Gates' kludged-together operating system probably seemed (and still does seem) preferable to many people. I am not competent to make a judgement call on this or to evaluate other alternatives (for example, other operating systems such as Linux) nor are such evaluations relevant here. What is relevant is the emphasis on the development of a new creative tool that is both in itself beautiful and elegant and highly functional *and also flexible*. Steve Jobs deserves credit for formulating this goal. Apple, at present, seems to be making the most progress toward this goal, and it has been adopted, at least in principle, by most of its competitors. It is undoubtedly the way of the future. Steve Jobs creative genius was not purely in art nor science but rather in the foresight to creatively combine them. He understood both were of vital importance in designing this latest creative tool: the computer.

Steve Jobs died in 2011 after amazing success in promoting his vision of what the new technology should be like.

MONKEYS AT THEIR WORD PROCESSORS

"Machines take me by surprise with great frequency."
—Alan Turing (attributed)
.

"Inspiration is easy; it's evaluation that is difficult."
—Hippokrites

The poetry reading was being held in an art gallery, and as usual the audience made up in diversity for what it lacked in size; everything from blue-haired matrons to black-haired 'gothic' teens. The featured reader was introduced as one of the most original of the current crop of avant-garde poets. He perched himself on a stool and picked up a book which he seemed to open at random. I was sitting close enough to see the title and author of the book: it was *The Sound and the Fury* by William Faulkner. Now not only did this poet not look like the photos of William Faulkner I'd seen, he didn't look dead, and I was virtually sure Faulkner had shuffled off this mortal coil back in the nineteen-sixties. The poet began to read quite dramatically, turning pages at the rate one would only expect of a graduate of The Evelyn Wood Speed Reading School. I couldn't find any semantic thread to follow, but that isn't unusual in hearing a complex contemporary poem read aloud without having the text with which to follow along. The rhythm was not regular, but again that isn't unusual in modern verse, and there were occasional brief interesting metrical riffs. Also occasionally a startling image or fascinating phrase would surface from the torrent of words. I quite enjoyed the first ten minutes or so of this reading, but unfortunately the poet read for almost forty-five minutes, by which time my Attention Deficit Disorder was making me really twitchy.
.

What was this poet up to? I found out that he was reading the first word on each line of text in the novel. I had to chuckle, for *The Sound and the Fury* is difficult enough to follow if you read every word. Similar techniques for creating 'found poetry' have been around for some time, so I don't think he should've been described as being all that avant-garde. And of course I can't imagine buying any of his books, assuming he had any, anymore than I can imagine purchasing a CD recording of Cage's *4'33"*—and, besides, I already own this novel by Faulkner.

.

This is not to say the reading wasn't an interesting experiment, but this is didactic art, and as a general rule didactic art isn't very good art. However, I did learn or relearn something from the reading, so it succeeded in its modest intention. What was the lesson, you ask? Well…

SIMIANS, SIMULATIONS AND ITERATIONS

It is unclear who first said that enough monkeys each punching keys on a typewriter will *eventually* produce the complete works of Shakespeare, but it—or variations on it—have been repeated many, many times. Usually the intent of the reference is to point out that even seemingly improbable events (for example, the development of a planetary system or the development of life from the primordial soup) are bound to occur given enough time. Sometimes, however, the point is quite the opposite—that something is ridiculously unlikely.*

What is indisputable, however, is that given enough time or enough monkeys (which is to say, a large enough number of seconds or simians), the Bard's sonnets are going to come rolling out of one of the typewriters. And what is of relevance here is the aesthetic implications of this fact.

In May 1989 the Dilbert comic strip (by Scott Adams) played on this theme. Dilbert writes a poem and presents it to Dogbert. Dogbert says, "I once read that given infinite time, a thousand monkeys with typewriters would eventually write the complete works of Shakespeare." Dilbert, impatient, asks, "Yes, but what about my poem?" To which Dogbert replies, "Three monkeys, ten minutes."

So be the work in question a Shakespearean sonnet, or Dilbert doggerel, the disconcerting thing about this monkeys-at-typewriters idea is that it, at least theoretically, removes the need for a sentient artist. We can, of course, find ample solace in the obvious unlikelihood of random keystrokes producing literary masterpieces. Or at least we *once* could—until recent advancements in computer technology gave us giga-monkeys, super-secretaries, capable of typing trillions of words a minute.

---

* Eddington is an example. "If an army of monkeys were strumming on typewriters they might write all the books in the British Museum. The chance of their doing so is decidedly more favourable than the chance of the molecules returning to one half of the vessel." (In *The Nature of the Physical World*)

AI AND AC: ARTIFICIAL INTELLIGENCE AND ARTIFICIAL CREATIVITY

I've often wondered about the phrase "Artificial Intelligence". Isn't all intelligence artificial? It certainly doesn't seem to come naturally to us humans.

.

There are two schools of AI—the weak and the strong. The strong AI-ers believe that it will eventually be possible to develop machines that are intelligent in exactly the same way we are, but even more intelligent than we. The number of such true believers has decreased somewhat since the heady early days of AI research, because— despite incredible, unexpected increases in computing power—really very little progress has been made on the road to that long-term goal.

.

But first it is worthwhile to look at the weak AI-ers—or rather to try and put the very idea of artificial intelligence in perspective. One way to do that is to think about AA—Artificial Athleticism. Human beings love physical competition. Consider something as simple as a foot race. I'm not sure who currently holds the record for "fastest man in the world", but I could beat him easily using AA. Let him get into his brand new Adidas running shoes, while I get into my 25-year-old VW Camper Van. But even though I'd win the race, I don't think it would affect my competitor's self-esteem. He wouldn't feel his worth threatened or feel that his long hours of arduous training were wasted.

.

A motorized vehicle isn't a track star in the same way that a calculator isn't a mathematical genius. Even a Paul Erdös, or a John von Neumann couldn't compete with my $9.95 calculator in a race to find the square root of—say—3,985,762. This is not to downplay the significance of the development of these AA and AI devices. If I need a package delivered to San Francisco, I'm not going to send a runner. Such foot-messengers have been made obsolete, replaced by a machine. And if I need to do a complicated calculation to get money back on my income tax, I'm not going to hire some fellow who is good at doing sums.* We have adjusted quite well to the fact that machines can now do many physical and mental tasks far better and faster than can we mere humans. But although the artificial

---

* Which incidentally is the original meaning of computer—a *person* who does computations for a living.

'intelligence' of a calculator is useful, it doesn't really impress us with its intelligence—and so we don't see it as threatening.

.

But let's move up a step in our demands of artificial whatever by returning to the overlapping ideas of intelligence and creativity. What was impressive about great mathematicians such as von Neumann or Erdös was not their ability to do sums in their heads; it was their ability to reason, to pose problems and then find solutions to them— in short to be creative. Can a computer be creative? Is there such a thing as AC—artificial creativity? Can a computer write a good poem, or even a single decent line?

.

The monkeys at their typewriters are profoundly stupid. Let's replace them with computer programs, equally stupid, but more capable of being wised up. Instead of monkeys pressing keys, let's have the computer's random number generator press the keys. And then, let's strive to increase the probability of creating a coherent, grammatically sound, English sentence. (We'll worry about meaning or aesthetic value later.) The starting point is a monkey or computer program that randomly types or generates a single character and then another— and another. One could quibble about how many characters to randomly sample from, but certainly less than a hundred would cover all 26 letters in both upper and lower case, blank spaces, and all conventional punctuation.* This is more than the number of keys on a conventional typewriter. People with too much time on their hands have figured out how long it would take to create even a single real word, not to mention a sentence or *Hamlet*, by this method.†

.

So in its dumbest version the program just spews out these characters one at a time, and it is going to take a very long time indeed before something syntactically sound—never mind meaningful—is generated. But a computer can have a 'virtual keyboard" much larger than a real one, for computer programs have very large hands with many more digits than we or monkeys do. Let's give it a keyboard where every key is a real English word. Now every keystroke

---

* Computer text files rely on 128 numbers to represent all the letters, punctuation, indicators of new lines, as well as a whack of special commands such as tab-over. (This is known as the lower ASCII, equivalent to half a byte of digital information.)

† I've found several Internet sites that do the math. I've also found one that does the typing: it has an amusing header that requests immediate email from any site visitors who happen to catch the program typing a Shakespearean sonnet.

produces a real word. The chances of producing a coherent sentence have just increased dramatically.

.

However, we'll still get a lot of 'sentences' such as "Dog windmill the ate computer haberdashery madly fun." But there is a way to further increase—even virtually guarantee—the production of grammatically sound sentences. Two things define every sentence: they are the sentence structure (or syntax) and the actual words. Consider this sentence: "Frankie bit the dog!" If a computer programmer codes the syntax of this sentence using a 1 for a proper name, 2 for a transitive verb, 3 for the definite article, 4 for a noun, and 5 for the exclamation mark—then the code for the syntax of this sentence is 12345. The sentence "Joan loved the pastrami!" has the identical syntax; all that is different is the actual words (of the correct part of speech) that are plugged into the syntax. A coded syntax of 34215 with the same words as the first example would yield this sentence: "The dog bit Frankie!" It would also fit sentences such as "The computer hates Ken!"

.

Legitimate, grammatical, syntax is defined as sequencing of parts of speech according to a set of rules particular to the specific language. Try some of the other variations of the simple coding above. Any sequence with the code 5 (exclamation mark) in a position other than last would never occur in English: "! The dog ran." is not a grammatically 'legal' sentence. Similarly any code starting with 2 (transitive verb) would be unacceptable in English without some auxiliary punctuation: e.g., "bit Frankie the dog!" is a disallowed syntax in English. There are a finite number of legitimate syntactical structures. If a computer programmer puts these, and only these, legitimate syntactical structures into a database and puts all the legitimate English words into another database sorted by parts of speech, then the generation of real, grammatically correct sentences is a simple mechanical process. First the computer randomly selects a legitimate syntax. Then it randomly selects each word for that syntactical structure from the appropriate part-of-speech category in the word database.

.

Many computer programmers have played with this idea, sometimes called a doggerel generator.* The first thing one notices in running

---

* It happens to be what first got me interested in programming, and I have written a moderately sophisticated piece of software called "Ghost Writer" which I freely distributed on the Internet.

such a program, is the surprisingly large number of sentences that are actually meaningful. (Another thing many people notice is how many sentences seem to have sexual innuendo, but that is another issue— probably related to the Freudian concept of projection.)

.

Now imagine running this program on a high-speed computer with a subroutine that checks every fourteen lines of output for a match to one of Shakespeare's sonnets. I don't have enough time on my hands (or the math skills) to estimate how long it would take for the program to 'write' a Shakespearean sonnet, but I can say with confidence that it is a very, very small fraction of the time the monkeys would take—and a small enough time period to give one pause.

.

Even more disconcerting is the fact that such a program often produces good lines of *original* verse. When I've run my own *Ghost Writer* program on a desktop computer at get-togethers at my house, it tends to be the life of the party.* People gather around the monitor exclaiming with great frequency "God, that's quite poetic." It's enough to give a human poet an inferiority complex!

---

* I know: this may say more about my ability as a host than my program's ingenuity.

REDEFINING INSIGHT: SIGHT INTO WHAT?

There is a profound lesson to be learned here, but let's backtrack to the question of artificial intelligence. I've said some minimal level of intelligence (by its current, popular, albeit limited definition) is prerequisite to creativity. If the strong AI-ers are wrong and computers can never attain true intelligence, then clearly they can not attain creativity.

.

Here is a personal anecdote which indicates two somewhat contradictory things about the mug's game of predicting what is possible: 1) it is prudent never to say never; and 2) one should be cautious about what constitutes a proof of something,

.

I once was a mediocre tournament chess player, but that still put me way above the average casual player. Like most chess players, I tended to view the game as a battle of pure intelligence and creativity, unsullied by any chance elements. If you won, you were smarter than your opponent; if you lost, your ego had to be bruised, for you couldn't just claim you were dealt a bad hand. In one of the first professional tournaments I entered I was amused to find that one of the other 'players' was a computer program. Word down the grapevine was that the program didn't even know how to castle! Needless to say it lost all six of its games. But a few years later, another computer program (one that had mastered all the rules) actually won one of its games—much to the chagrin of its opponent. Now a chess program that has learned all the rules is like a writing program that has learned to only write grammatically sound sentences, but a computer program beating a human opponent at chess is like a computer program winning a writing contest. I believed that chess was an art form, and so this victory of silicon over grey matter unsettled me. Nevertheless, the particular human the computer defeated was what chess players called a 'fish', a less than stellar player entered in the lowest rated section (derisively call the aquarium), so I made pronouncements to anyone who would listen about how no computer program would ever beat even a B-Class tournament player such as myself. Chess required intelligence, so went my reasoning, and computers could never be truly intelligent, despite the claims of the strong AI school of thought. Of course it follows from this that a computer that plays brilliantly, that wins against a brilliant chess player, proves the strong AI-ers were right.

.

My naiveté and hubris blinded me to the fact that it was incredibly unlikely that a computer which had defeated even a weak player could have won by making random legal moves: it would had to have 'made judgements'—and that once a computer can be taught to make even half-decent judgments, it can be taught to make better ones.

The rest is history—and my own words served up for lunch. Even those who don't know the rules of chess know that a computer program called Big Blue beat the best human chess player on the planet. Closer to home, I know that the chess program I have installed on my desktop does me in—even when set at a humiliating low skill level.

How to save face? Decide that chess isn't really an intelligent activity and certainly not a creative one? Well, of course this depends on what one chooses to call intelligent or creative.

I could fall back on the argument that the program didn't win: the programmers did (which included several chess experts)? However a hole is quickly punched in this argument by pointing out that none of the programmers of Big Blue could have defeated the world champion. Yes, it took great skill to write this program, but it takes great skill to raise and educate a child. If the child's skill eventually far surpasses the parent, one can't denigrate its intelligence and pass the credit back to the parent. (Although some parents try to do this.)

Can one claim it was all mechanical evaluation, gears whirring (actually electrons flying), and it only won by the sheer brute force of calculation? Unfortunately, this isn't actually true, only a common misconception about how computer chess programs work. I know enough about programming to know that this is not how a chess program works. It doesn't just analyze all the possible outcomes of all possible moves in a particular situation, for even with our best super-computers, that is impossible. Chess programs are programmed to make judgment calls based on rules. Like humans, they first eliminate some moves as probably silly or useless. Then they analyze the remaining moves some 'plies' deep and compare the resulting positions according to guidelines as to their value. They can analyze more moves, and analyze these moves deeper, than the human brain can in the same period of time. But that is their only advantage. Their disadvantage is the inflexibility of their evaluative rules.

Well then, at least one can deny the program artificial intelligence, because it *wasn't conscious of winning*—although it did announce its victory. Well, this can't actually be tested, but it seems reasonable to assume my desktop chess program isn't sentient—something that offers some salve to my ego. Unlike too many of my human opponents over the years, it doesn't gloat at its victory and ask me condescendingly "Where d'ya think you made your *big* mistake?" (But of course it could be programmed to say that, and some arcade games are programmed to make derisive remarks when the player is 'killed'.)

But this brings us to the central issue in the debate about artificial intelligence and creativity. What criterion can we apply to determine if a mechanical thing has the right to be called 'intelligent' or 'creative'? We can't get inside another person's head, so we can't really know if the bank teller or our spouse is truly conscious, sentient, *intelligent*. (Sometimes I'm sure we all have fleeting doubts.) We have to rely on observable things, on phenomena—and make what we hope are justified assumptions. I can't tell if my students have learned the material in one of my courses. I can't measure learning directly, because I can't open up their craniums and look inside for the content I hope they have stored there. I can only give them a test, measure their performance, and assume (hope, pray) it reflects what is in their heads. In doing so I am making a very big assumption, making a huge leap of faith.

The risk of 'false negatives' on exams is very large. Joe knows his stuff, but he comes to the exam with a hangover, or a bad case of the jitters, or a flu virus. He fails, and it's his tough luck. False positives are less likely, but I'm sure some students 'luck out' in getting an exam that just happens to cover only material they studied. And then of course there are those successful cheaters who score well, but know nothing.

So how do we test for intelligence in a machine, since we obviously can't get inside its 'head'?
The classical answer was offered by Alan Turing and has come to be known as the Turing Test. The scenario runs like this. There are two sealed rooms. In one of these rooms is a human being sitting at a computer terminal. In the other room there is a computer with an AI computer program running. The examiner is outside these rooms, seated at his own terminal. He types in questions sent to both rooms and examines and compares the anonymous replies he receives from both. If he is unable to consistently ascertain which room contains a

human being and which a computer, the computer is deemed 'intelligent'. This is clearly a very solid operational definition. We only know our spouse is really intelligent, and not just a mindless android constructed of some very new flesh-like plastic, because of what she or he says in response to what we say and do.*

Within the AI community, this test of sentience is more or less taken for granted as the standard, which is not to say it hasn't been strongly criticized by certain philosophers, most notably John Searle with his famous Chinese Room Refutation. As computers increased in power and influence in our world, universities everywhere opened departments of "Cognitive Science" that were committed to a hybrid discipline of computer science and psychology. These departments required of their faculty an oath of allegiance to the idea that the brain is like a digital computer. And getting tenure often required becoming a card-carrying member of the strong AI-ers. When John Searle presented a simple logical refutation of the Turing Test, he caused serious depletion of our natural forests: paper (once trees) flew everywhere, as books and journal publications attempting to refute his refutation came spewing out from the threatened Cognitive Science departments.

Like C. P. Snow before him, Searle is often grossly misunderstood and demonized as the enemy of what he actually supports. Snow wanted to bridge the two cultures and felt that acknowledging its existence was the first step toward reconciliation. Searle wants to understand intelligence and thinks that clarifying what is meant by it, and what is a true test of it, is an essential first step in the process.

His Chinese Room Refutation came to him while sitting on an airplane en route to a lecture he'd agree to give to some cognitive scientists. It is a very simple and elegant argument. Assume the Turing Test is for intelligent discourse in Chinese. Imagine that in one room is a native speaker of Chinese, and in the other room, instead of a computer, is a man who knows no Chinese, but is equipped with a rule book that indicates what ideograms to send back in response to the ideograms sent in. The interrogator sends a question to both rooms. It doesn't really matter what the question is as long as an intelligible answer is possible. The someone fluent in Chinese reads the ideograms, understands the question, and replies.

---

* This is a good place to diplomatically refrain from obvious wisecracks.

The other guy looks up the ideograms in his rule book and sees what ideograms to send as a reply—but is blissfully unaware of both the question and the answer he sent. The interrogator can't detect any difference in the replies he gets from the two rooms and concludes both contain people who understand Chinese. The Turing Test has been passed by some guy who actually has no understanding of Chinese!

.

My own heretical hunch is that Turing was not so much offering an objective test of intelligence or sentience as satirizing excessive faith in operational definitions, such as those so common in Behaviourist Psychology. But whatever his intention, the fact is that some computer programs have already passed the Turing test—but only programs whose operation is so transparent to anyone with the simplest programming skills that it makes a mockery of the idea that the program is 'intelligent'. The critic of strong AI, Joseph Weizenbaum at MIT, developed one such program back in 1967. The program pretends to be a clinical psychologist of the non-directive school. The 'client' talks to this 'therapist' by typing on a computer keyboard, and the program replies with comments and questions presented on the monitor. The story goes that one of the secretaries in Weisenbaum's office was playing with the program, which she *had been told* was a computer simulation, and after half an hour of interaction with the program became convinced it wasn't really a program at all, but rather that it was computer terminal hooked up to a real person somewhere else on campus!

.

Without going into details, the program works by storing certain keywords the client has typed (such as Proper names or emotionally charged words such as mother or sex) in a database. It then, sometimes quite some time later, uses something like the doggerel generator previously described to write sentences and questions using these words. For example if the client types in a person's name, say Jane, then some time later the program will generate a sentence such as "Let's talk a little more about Jane. How do you really feel about her?" Anyone who can read whatever computer language Eliza is written in can see how totally mechanical this process is.

'

If Searle needed any empirical support for this logical refutation, this charming program not only demolishes any claims to the validity of the Turing Test, but also pokes a bit of fun at non-directive psychotherapy. Sure, one might argue that the secretary who became convinced this was a real therapist wasn't the sharpest knife in the drawer—or that the story is apocryphal.* However, using an algorithm called CBR, or case based reasoning, the program has been greatly refined, and anyone who tries the latest version will have to admit that it could fool at least some people.† And although no one has yet developed a program that can consistently pass the Turing test if a sceptical scientist takes it, the program makes an irrefutable point.‡

.

So, assuming the Turing Test isn't the final test of artificial intelligence or creativity, what is? What would convince anyone that a computer was being truly creative?

---

* Or, a cynic might suggest, that she'd had some previous experience with a real non-directive therapist.

† The program is available as freeware on the Internet.

‡ Programs that talk with humans can still be easily tricked if they are asked questions about commonplace knowledge, such as the color of bananas. The database of knowledge, even the most uneducated have in their heads far surpasses any computer database developed. But that is only a matter of quantity. It is absurd to assume that when enough information is stored in a computer database, the computer will suddenly become sentient.

REDEFINING INSPIRATION: TAKE A DEEP BREATH

There are computer programs that write verse: I've even written one, already mentioned, called *Ghost Writer*. There are computer programs that write fiction—or academic articles in the field of hermeneutics*. There are computer programs to generate images, both abstract and landscape. There are computer programs to generate musical compositions. One of these programs is called EMI (Experiments in Musical Intelligence), written by the composer David Cope. In response to having the composer's equivalent to writer's block, he went to his computer and programmed it to analyze the works of the musical greats, including Bach and Mozart, looking for recurring patterns, in all the parameters that define sound, within the compositions of specific composers. Then he synthesized some of these patterns and had the computer play them. To his amazement in the musical pieces so generated he could hear and distinguish the 'ghosts' of every composer he tried this with. Cope's continued experimentation with EMI has resulted in numerous concert performances and half a dozen CD recordings, including ones of "Virtual Bach" and "Virtual Mozart. Of course he tweaks the computer output, but the fact remains that the human performance of these AC compositions could fool most Bach and Mozart fans into believing they were hearing a newly discovered work by these musical giants.

.

It seems to be getting more and more difficult to find a way to save the dignity and uniqueness of the human creator. Is the battering my inflated ego took, when those chess programs I once mocked began kicking my ass on the 64 squares, destined to be repeated in my literary and artistic endeavours? In all artistic endeavours, even music? Cope tells of actually being physically assaulted by an outraged and indignant composer at a conference where one of these computer compositions was presented.

.

In the summer of 2007 I gave a talk at a conference entitled "Human Mind - Human Kind". The theme of the conference was "What Are The Uniquely Human Behaviours, Social Practices, And Psychological Structures That Make Man Particularly Human?" Is there anything we can claim as unique to us as *Homo sapiens*? I somewhat glibly dismissed the usual suspects: language, tool-use,

---

* This is a rather easier task.

transgenerational transmission of culture, cognitive skills at problem solving. There is a least some evidence (dependent on how stringently one operationally defines them) of these characteristics in other species, albeit not as strongly manifested as in humans. The last man standing seemed to be the creativity that is the foundation of art and science on which civilization is built. In my talk I considered the claims that elephants and chimps can create primitive visual art, but more to the point here is the question whether or not computer program entities, such EMI as composer and Ghost Writer as writer, can create art, be creative in the same sense that human beings are creative.

.

For the moment, put aside philosophical questions about computers ever becoming sentient entities—or conscious. What does that matter, if they can out perform you in thinking *and* creating, even if they are nothing more than Turing Machines following the detailed instruction sets called algorithms? Does it ease the sting if your competitor is insentient, even unconscious, as it consistently out-performs you?

.

I think it should! "O Death, Where is thy sting?" Death is natural and is not sentient. There is no sting. The creation of improvements on our own biological facilities is natural to human beings. There is no sting. My sense of worth as fast bicyclist is not diminished by the fact that a car can beat me in a race. No mathematician feels his self-esteem destroyed by a calculator. Consciousness and what philosophers call 'intentionality' matters.

.

However, before addressing this very important issue, it is valuable to consider the source of 'artificial creativity'. A computer simply follows instructions. So how on earth is it possible for it to create anything new and unexpected at all? The program determines, in the most deterministic sense possible, the outcome. You can program it to calculate wind-chill based on temperature and wind speed, and it will do it faster than you can, but the outcome is determined—and can be confirmed by—laborious human calculation. Like getting a room at a Holiday Inn (at least according to their earlier ads), you will have no surprises.

.

Yet creativity is about surprises!

.

Well, one thing that creates surprises is chance, a random event. I used to tell my computer science students that a computer does have a soul, a creative centre, and it is the random number generator. Insert one line of code that calls up the random number generator function, and your program output will be different, unexpected and unpredictable most every time.

.

The random number generator in a computer does not generate a truly random number, for only quantum decay can do that, but if seeded with the computer clock, it is random for all practical purposes. And the number it generates can be used to select random words to insert in randomly selected syntactical structures, and so produce entirely unpredictable sentences or lines of verse.

.

A computer program's 'inspiration' consists of random events inserted in the code. And, as a poet, I have to admit that apparently random events in my head are the usual initiation of the scribbling which eventually becomes a poem. (I've already discussed the role serendipity plays in scientific discovery.)

.

So is creative inspiration no more than fortuitous output 'created' by a computer's random number generator or the inexplicable firings of neurons in the creative person's brain? The answer to that question is surprising, for it may be that creating something new is a trick even a moronic computer can do, but recognizing it as a good trick is the real creative accomplishment.

THE EASE OF CREATION AND DIFFICULTY OF EVALUATION

One of the major, and largely unforeseen, roadblocks in the development of AI has been at the beginning of the circuit. We don't have robot servants doing our housework, as predicted twenty or thirty years ago, not so much because we can't program them to 'think' or move, but rather because we can't program them to see! Teaching computers to *recognize* even the simplest objects has turned out to be extremely difficult.

.

The problem in the development of AI and AC is at the evaluative end, not the creative beginning. A computer randomly generating text will occasionally write a good line of poetry, so it is as good as a poet in that way. But it, unlike the poet, can't *recognize* that it is a good line.

.

This has profound implications for our understanding of creativity: *What may very well be the defining characteristic of creativity is not the generation of ideas, hypotheses, images, poems, or whatever—but rather the ability to recognize what is worthwhile.* The poet rewrites each line again and again and again until he recognizes it as right or as good as it is going to get. Artists may do a hundred sketches before recognizing one as good. Scientists may consider many hypotheses before recognizing one as potential fruitful.

.

Of course they all can and do make mistakes. Lines of verse can leave the reader unmoved, sketches leave the viewer bored, and hypotheses fail to be supported by empirical evidence. But the measure of creative accomplishment isn't the number of failures, but the number and importance of the successes. As Vince Lombardi is credited with saying: "It isn't how many times you get knocked down, it's how many times you get back up." As previously remarked, highly creative people are highly productive people. Goethe 'wasted' his time on faulty pseudo-scientific studies. Newton 'wasted' his time on his efforts in alchemy and theology.

.

It is humbling to think of a writer, for example, as a doggerel generator with an efficient algorithm, to think of the poet's Muse as a random number/word generator. But if the critical factor in productive creativity is the ability to recognize what is worthwhile, creative people have little to fear from computer technology. If, as I have argued, the key to understanding, defining, evaluating *and creating* both art and science is the aesthetic experience, then computers will only become truly creative when they become capable of having aesthetic experiences. Some strong AI-ers may believe that will eventually happen, but I, for one, wouldn't bet on it. And if it does happen, it will mean we mere mortals will have been displaced from our position at the top of the evolutionary hierarchy and the responsibility of continuing civilization passed to silicon life forms. But like Searle, I think this logically impossible unless we devise something that can *truly* understand the art we create.

.

This is not to say that computers won't be able to make some damn good judgment calls without appreciating they are good. As already pointed out chess programs already do just that, and do it well enough to defeat humans at their own game. Expert system programs such as the medical diagnostic program MYCIN, behave much like scientists, generating reasonable hypotheses and testing them against known data. There is no reason a program to generate images can't be instructed to reject images containing colours that are known to clash or that fail some other established aesthetic principle. Still, such filters, being exclusive by definition, cannot replace the particularly human ability to have an aesthetic response. Most great, innovative artists and scientists broke the currently accepted rules—precisely those rules that a sophisticated programmer would use as filters to reject a randomly generated scientific hypothesis or work of art. How many visual artists have used clashing colours to great advantage? How many poets have shattered the rules for correct syntax and so moved the reader in a truly new way? How many scientists have advanced knowledge by testing hypotheses absolutely contrary to the currently dominant and accepted paradigm? One characteristic of the truly creative is that the mental filter used by their less inspired contemporaries is 'broken'. *They are not imprisoned in algorithms.*

.

So although the new computer technology seems unlikely to replace the creative human, it can *contribute* to the productivity of that human. As the examples given in the preceding chapters clearly indicate, artists and scientists possess new and powerful tools never before dreamt of in anyone's philosophy. And one might even say that AI and AC programs might even be considered more than mere tools: they may even become kind of a new—Muse.

CASE STUDIES: ALAN TURING, GOD

I suspect my choice of the two subjects for his final case study will be startling even to the reader by now accustomed to some unusual pairings. However, in the context of this final section, I think it is a quite reasonable pairing, because both subjects offer powerful insights into the fundamental nature of creativity above and beyond our purely human version.

.

And remember I already set a precedent in these case studies by treating the fictional character of Frankenstein's monster as a legitimate subject, for he has metaphorical reality. The same is true of God. Since God has seniority, I'll give Him the last word and first look at Alan Turing, the troubled genius who was one the earliest explorers of this new territory of 'artificial' intelligence and creativity—and the one who was sentient and conscious.

.

Alan Mathison Turing was born on June 23, 1912 in London, shortly after his parents had returned from postings in the Indian civil service. He and his older brother were sometimes left with friends when his parents returned to India for duties. By now in this book, the reader should recognize this pattern of his early life and education as familiar. His precocity was recognized early and yet to some extent discouraged—which probably had the opposite effect. His secondary school (Sherborne in Dorset) was what the British refer to as a "public school"; such schools are in fact exclusive and expensive private schools with curricula designed to deliver a 'classical education' appropriate for the privileged classes that can afford them. His interest and aptitude for math and science was not considered appropriate, and the headmaster wrote his parents warning of the inappropriateness of this for a 'proper' liberal education.

.

Turing was homosexual. While at Sherborne he fell in love with his friend, Christopher Morcom, who unfortunately did not feel the same way toward Turing. When Morcom died suddenly from tuberculosis contracted from drinking contaminated milk, Turing was devastated. Homosexuality, far more than even interest in science and math, was not socially acceptable at his school or in his social class, and Turing had earned early to keep his feelings to himself.

.

He then attended King's College at Cambridge where he distinguished himself. Shortly after graduation in 1934, he was elected a 'fellow' of the college on the basis of his dissertation on the Gaussian error function.

In 1936 he published the classic paper "On Computable Numbers, with an Application to the *Entscheidungsproblem*", which described what was to come to be called "The Turing Machine". It changed the face of mathematics and logic forever and established algorithms as the solution to any conceivable mathematical problem. This is the cornerstone of the whole science of computing, and it is for this reason that Turing is usually honoured with the title of "Founder of Computer Science". Of course, there were no real computers at the time (the technology wasn't to develop until many decades later), but the principles he discovered have been demonstrated in practical applications even he could not have imagined.

At the outbreak of World War II, he applied his genius to the practical and important task of code breaking. He developed "The Bombe", an electromechanical device which could find the settings for the famous "Enigma Machine" used by the Nazis to code their messages. He is justifiably considered a war hero for his contributions to the Allied Effort; knowing what the enemy knew and was going to do was critical. During the Normandy Invasion, it is estimated that as many as 18,000 German messages were being decoded.

When the war ended, Turing took a position at the National Physical Laboratory outside of London and created the design for a computing device that could store algorithms; i.e., computer programs. In 1947 he moved to the University of Manchester to work on the development of one of the first real functional computers, the Manchester Mark I.

Until his death in 1954, he continued to make creative contributions in such diverse domains as mathematical modeling of biological events[*], the philosophy of mathematics and the emerging science of computing. His Turing Test, already discussed, is a seminal philosophical thought experiment about the then non-existent domain of artificial intelligence. As is the case for all of the subjects

---

[*] For example, Turing published a paper on the subject called "The Chemical Basis of Morphogenesis" in 1952, putting forth the Turing hypothesis of pattern formation based on the Fibonacci number sequence that appears in plants, such as the sunflower.

of the case studies in this book, this brief summary of his contributions to civilization is appallingly inadequate.*

The cause of Turing's death remains controversial. Homosexuality was still illegal in England in the fifties, and when in 1952 Turing reported a break-in to his house by a young man who it became evident was probably a disgruntled lover. Turing (obviously really the victim) was charged with "gross indecency", and his governmental security clearance was revoked. He was summarily convicted and given the choice of either imprisonment or a probation granted on the condition that he would agree to undergo hormone injections to reduce his sex drive. He opted for the latter and received oestrogen injections that caused breast enlargement and, undoubtedly, profound emotional distress.

In 1954 he was found dead by his house-keeper. A half-eaten, cyanide-laced apple was found by his body. While the official, and plausible, consensus seems to be suicide as the cause of death, there are those who believe (e.g., his mother) that it was accidental because of his careless handling of dangerous chemicals, and others who believe he was actually assassinated because he posed a security risk to the government!

Alan Turing's life, like so many creative lives I've so briefly summarized in these case studies, is tragic. Sometimes it seems the cause of the tragedy is to be found within the character of the individual and may (sad to say) be inevitably entangled with the person's creativity. Sometimes it seems that the cause is external: a world that cannot accept deviance even though it is deviance that moves us as a sentient species forward. Yet deviance that is the kernel at the heart of creativity.

I would venture to say that Alan Turing's most monumental contribution to the understanding of creativity is his exploration of the role insentient and unconscious, but *not* random, pattern development plays in everything. And how we can, once cognizant of this, use it to creative advantage.

---

* I hope I can be forgiven, for were I to do justice to the accomplishments of the many creative individuals I've chosen to use to make a point, this wouldn't be a book: it would be at least a 20 volume set of biographies—which I'm certainly not qualified to write.

Now I shall wind up this final case study with a look at the greatest creative genius ever, even though there is no evidence whatsoever that He is sentient or even conscious. Yup, I'm talking about the big guy: 'God'. His parents were Mother Nature and Father Time, but—somewhat paradoxically—he was conceived by *Homo sapiens* only about 200,000 years ago.

.

By now everyone has heard the Creationist argument that if you find a volume of Shakespearean sonnets on the beach, you must assume it couldn't have come into existence by chance: it had to be written by someone. If you find a watch in the sand, you have to assume someone designed and made it. It was when the brains of *Homo sapiens* evolved to the extent that they could wonder about the source of what seems a clearly designed world inhabited by creatures that surely couldn't have happened by chance, be they gazelles or themselves, that God was conceived.

.

Actually, and paradoxically, God existed long before this flawed and naïve human conception of Him. As far as we can tell, His parents Mother Nature and Father Time were born simultaneously and spontaneously approximately 13.7 billion years ago in what is called "The Big Bang".* And of course neonatal God promptly sprang into existence as well and set to work.

.

Remember the monkeys at their typewriters. Mathematicians, with too much time on their hands, will tell you that you'd need more monkeys than ever existed and much more than 14 billion years to produce that volume of Shakespearean sonnets. This isn't an argument for God, The Great Designer, as conceived of by religious fundamentalists. This is an argument against the idea that what 'God' does is purely random. This is precisely what the Creationists fail to realize. For example, as numerous writers such as Dawkins and Dennett have tried to point out, evolution by natural selection is *not* random. Natural selection is the brilliant explanation of how the wonders that comprise our wondrous world can result from non-random events without any sentient help, without any guiding hand. God, in the metaphorical sense I am using that word (the same sense that Einstein used it), *consists of* the marvellous processes of nature operating through time.

.

---

* Why has no one I know of has ever commented on the colloquial sexual connotations of that phrase? Is it just my dirty mind?

Remember the monkeys at their typewriters. If all the keys pressed produce legitimate English words or phrases used by Shakespeare (rather than just letters), creating those sonnets is no longer entirely random—and will take much less time, albeit still a very long time. If the algorithm in a *Ghost Writer* computer program is refined to only select among legitimate syntactical structures, you are only going to get real sentences. Sure, only some of these grammatically correct sentences will make semantic sense, but that is true of such natural events as occur in evolution. Only some changes will make sense, but the principle (algorithm) of natural selection assures that those that don't make sense are eventually discarded.

The fundamental laws of physics work in the same algorithmic way. The keyboard being pressed by random quantum events isn't entirely random: it is limited to things that 'make sense' by these laws of nature. It still takes a very long time and a lot of space (both of which were available) to create the universe we inhabit, to create the outlier planet in the suburbs of an insignificant galaxy that we inhabit, *and* to create a species on this distant outpost that can make watches and write sonnets.

But it happens. Obviously. Because it has happened.

Human creativity *is* a miracle. It isn't an inexplicable random event, but it isn't supernatural either. This does not mean it isn't still miraculous. Bach remains a miracle, as does our ability to appreciate his music.

We don't need Him to explain things, but what the hell: God bless!

EPILOGUE: THE GIFT

"The Gift of Truth excels all other Gifts."
— Gautama Buddha

"Someone I loved once gave me a box full of darkness. It took me years to understand that this too, was a gift."
— Mary Oliver

Civilisation is defined by the accomplishments of the arts and sciences. And all those accomplishments are made possible by our creativity. I often call creativity our greatest 'gift', but unlike with most gifts, there is no one to thank. The Muses and all deities are mere myths invented to explain the wonder of creativity—and life itself. Like life, it is a wonder to be appreciated, just as any gift bestowed on us should be.

There is a saying that one should never look a gift horse in the mouth. This may simply mean that it is rude and inappropriate to promptly examine a gift for flaws. However, if it implies that examining a gift will devalue it somehow, then I strongly disagree. I believe that understanding is always desirable, an end in itself, even if it is disturbing. And understanding has to always increase our appreciation, wonder and awe.

Further understanding of the nature of creativity certainly does that. It also possibly might help in nurturing it and using it for good ends.

This is why it has so interested me. It is what motivated me to write these *Secret Agents* books.

AFTERWORDS

In other words, stuff stuffed at the end.

ACKNOWLEDGEMENTS

I think virtually everyone, upon beginning to read a book, skips the acknowledgements—everyone, that is, except those who feel they should be acknowledged. Better, it seems to me, is to do as they do in the movies: put the credits at the end.

.

I'm limiting my acknowledgements to those who have been so particularly helpful and supportive that they shouldn't be buried in a huge list that no one will read. However, I also would like to extend an all-purpose, less specific, thank you to some other folk as well.

.

There is no way I adequately thank my wife, Ursula, for all she has done for me. She has always been an indefatigable editor. Without her, I would've been continuously embarrassing myself in print from the very start. But most importantly, she has been the inspiration for all my work. She redeems my life. She is my Beatrice—and I got to actually marry her and not just admire her from a distance.

.

I also owe a debt of gratitude to our two 'kids', Christiaan and Katherine. They certainly deserve credit for their tolerance of their eccentric father, but, more importantly, for their passion for all the aesthetic and intellectual, and even purely physical pleasures, life has to offer. They have been an inspiration to me.

.

My family has taught me that loving another and nurturing that 'selfish gene' is the most creative thing we can do. And the most rewarding.

.

I also am incredibly grateful to the many scientists and artists whose creations were the ultimate inspiration for this book—and even for my way of life.

.

Thanks, too, to those several teachers I had long ago who through their own passion for art and science instilled in me the respect I have for creative endeavour, as well as to my students who have had to listen to me test out my ideas on them in what I'm sure were sometimes rather rambling lectures in my Psychology of Art class.

.

I should also acknowledge Nipissing University for giving this writer a very rewarding day job for 40 years. And much of the first draft of *Secret Agents* was written during a sabbatical leave.

SELECTED ANNOTATED BIBLIOGRAPHY

In my judgment the following works are well worth reading (or at least sampling) and *not* because I think they offer supporting evidence for my various theses. (Some do not.) I recommend them because I believe they are all entertaining and enlightening works that are at least tangentially related to the content of this book. Some are about creativity and creative individuals and some are the actual works of the creative individuals mentioned in *Agents*.

.

Recommending books is way too much fun, for who doesn't love plugging books and writers one admires. But I've tried not to over-indulge in this pleasure. Life is very short—even if one reads quickly—and so I've grouped my recommendations by the sections in *Agents*, so that if something along the reader's way triggers a particular interest, he or she can easily follow up on what inspired me.

.

I've almost always limited my recommendations to a maximum of six per chapter, usually with two of these referring to the subjects of the 'case studies' for that chapter. The list is still long, but rest assured—there will *not* be a quiz!

.

I admit that what I have chosen to include is extremely arbitrary and idiosyncratic and reflects my own reading rather than some omniscient overview of the many topics touched on in *Agents*. Anyone intimate with a particular chapter's theme could surely cite more relevant and better works, even though I've sincerely tried to select, given my own limited knowledge, excellent works I believe touch most of the bases of each chapter. I've included a very brief justification for the inclusion of each work in this list.

.

I've usually only given the author and title of the work, for in The Amazonian Internet Age that is sufficient. The only exceptions to this are when more information is really required to easily find the recommended text.

- **Going Where Be Dragons**
  - General suggestions and remarks.
    - o There are many silly folk still willing to play the mug's game of predicting the future. (I obviously am one of these silly folk, or I would've omitted this final pane of my triptych on creativity.) I would point the reader interested in such prophecies to magazines such as *Wired* or to good science journalism magazines or to the shelves of new science fiction at their local bookstore. The science facts about new developments spewing out everyday are a constant source of inspiration.
  - Recommended reading.
    - o <u>Gardner, Howard. *Five Minds For The Future*</u> – It was a hard choice to pick a single book that seemed both plausible in its predictions and sufficiently bold. This work by Gardner, a serious thinker about intelligence and creativity, seems to fit the bill, although I've always found his perspective a bit tinted 'rosy' and overly optimistic.
- **Sailing Over The Edge Of Inhibition**
  - General suggestions and remarks.
    - o The explicitly 'psychedelic' literature of the early seventies, like the 'psychedelic art' of the period, is mostly quite lame and uninteresting. However, the number of important writers who have used mind-altering drugs is far greater than the number who didn't—at least if one includes everything from excessive alcohol consumption to experimentation with the strongest narcotics or hallucinogens. Not surprisingly 'Drug lit' has become a major genre, and it is easy to find relevant material. Since alcohol has been the most commonly used mind-altering drug by writers, I've included in the following list the masterpiece of a novel by Malcolm Lowry that deals with the descent into alcoholism along with some works dealing with what are more often considered 'drugs'.
  - Recommended reading.
    - o <u>DeQuincey, Thomas. *Confessions of an Opium-eater*</u> – The author was associated with the Romantic

Movement in literature, and in this book he gives one of the first autobiographical descriptions of the descent into madness resulting from drug addiction. From his vivid descriptions of the initial drug-induced euphoria to the eventually nightmarish visions that drove him mad, this work is a disturbing and powerful classic.

- o Huxley, Aldous. *The Doors of Perception* – This is the Huxley's famous paean to drug-induced spiritual insight.
- o Leary, Timothy; Metzner, Ralph; and Alpert, Richard. *The Psychedelic Experience: A Manual Based on The Tibetan Book of the Dead* – This was the hippie psychedelic 'bible'—and instruction manual for good 'tripping' on hallucinogenic drugs.
- o Lowry, Malcolm. *Under The Volcano* – Published in 1947 after repeated rejections by publishers, this semi-autobiographical novel is widely considered one of the greatest novels of the 20th Century. It is analogous to—although far more complex than—DeQuincey's, *Confessions of an Opium-eater;* however here the drug of inspiration and destruction is alcohol.
- o Miller, Henry. *Time of the Assassins: A Study of Rimbaud* – This "study" of Rimbaud is more about Miller than Rimbaud, but since both figure so centrally in the new aesthetic of artist as the-one-who-leaps-over-the-wall-of-inhibition, it is worth reading.
- o Starkie, Enid. *Arthur Rimbaud* – While numerous factual details have been cast into doubt from subsequent scholarship, this book still stands as *the* defining portrait of one of the seminal figures to present "derangement of the senses" as the road to creation.
- o Wolfe, Tom. *The Electric Kool-Aid Acid Test* – This is Tom Wolfe's flamboyant, purple coloured prose description of the drug saturated adventures of the writer Ken Kesey and his infamous "Merry Pranksters". It truly captures the spirit of the sixties drug culture. At least I think so. (If you were there, you can't really be expected to remember.)

- **Strip Mining The Subconscious**

- General suggestions and remarks.
    - o Here I think it appropriate to make a suggestion regarding what *not* to read. There are an amazing number of books that claim to explain how to harness the putative power of one's subconscious to make one creative and generally fix everything that is wrong with one's life. Avoid them, even though they often have gushing testimonials from readers that are startlingly similar to those written by folk who 'have found Jesus'. (Strangely, I haven't discovered any of these testimonials from a notably creative person.) So for this section I'm simply including Osborn's and De Bono's books which outline their recipes for creativity and a critique of the cult of creativity and genius, plus the diary of the contemporary artist most noted for his claims to tapping the depths of the subconscious—and being a genius.
- Recommended reading.
    - o De Bono, Edward. *Lateral Thinking: Creativity Step By Step* – De Bono is prolific (and often tiresome and repetitious), but if one wants to examine his original and basic ideas, this was the book that made his reputation.
    - o Dali, Salvador. *Diary of a Genius* – This is a journey into Dali's labyrinthine mind, a mind as teeming with contradictions as are his surreal paintings. However, unlike his highly structured art, Dali's diary is a chaotic, albeit interesting, mess.
    - o Osborn, Alex. *Applied imagination: Principles and procedures of creative problem solving* – This is the book that started the brainstorming-can-make-everyone-creative craze. The important kernel of truth in his method is the emphasis on suppressing any criticism or censorship of ideas. If the subconscious is really dishing up ideas, then the 'sensible' conscious mind shouldn't be swatting them down as soon as they surface.
    - o Weisberg, Robert. *Creativity: Beyond the Myth of Genius* – Weisberg challenges all the major clichés about creativity and genius and argues that creativity is really no different from systematic problem solving and genius not some magical property of

rare individuals. One need not agree with all of Weisberg's conclusions to appreciate this unsentimental, hard-headed look at many of the implicit and dubious assumptions underlying much writing about creativity.

- **The Magic Of Juxtaposition**
  - General suggestions and remarks.
    - o I believe the success and universal appeal of surrealism is based on the magic of juxtaposition, which allows us to see both physical entities *and* ideas clearly because of their appearance in unlikely places. This re-colours them with the brilliant hues of emotion—and *emotion is what gives everything meaning*. In science, the juxtaposition of ideas from one sub-discipline to another, as with E.O. Wilson's study of insects and other species, fires off a similar flashbulb illuminating us in the grand landscape of all living creatures. I've kept my recommendations to only four, but they are all works I just cannot recommend too strongly.
  - Recommended reading.
    - o Alden, Todd. *The Essential Rene Magritte* – There are many worthwhile books on Magritte. This one contains full colour images and information about his life and work. Magritte may be the most important of the surrealists—for his work does not only force us to see physical objects with new eyes, it also makes us think about old ideas in a new light.
    - o Damasio, António. *Descartes' Error: Emotion, Reason and the Human Brain* – This is a book that should be required reading for anyone presuming to say anything about consciousness, emotion, or psychology. Damasio is a scientist with a poet's soul and a philosopher's mind. (His other books are also more than worth reading for anyone interested in the philosophy of mind and emotion.)
    - o Durozoi, Gerard (translated by Alison Anderson). *History of the Surrealist Movement* – This is considered the most up-to-date and comprehensive survey of surrealism and its influence on artistic genres all over the world.
    - o Wilson, E.O. *Sociobiology* – The last chapter of this book is what set off a firestorm of protest from

those who didn't like having their rosy-coloured glasses knocked off their smug faces—and so had to face up to the bald fact that we are just another species, a species in many ways totally driven by forces as purely biological as ants invading a picnic lunch.

- **The Idea Of Evolution; The Evolution Of Ideas**
  - General suggestions and remarks.
    - The implications of evolutionary theory (both in species development and in ideas) seem to remain difficult for many people to grasp and accept. Nevertheless, they are—and will continue to be—extremely important to the future of creative science *and* art. How could a better understanding of ourselves and our place in the universe not change the way we create our art and science?!
  - Recommended reading.
    - Brinton, Crane. *The Shaping of the Modern Mind* – I include this book (even though it is out of print and may be hard to find) just because it is my favourite brief book on the evolution of ideas and is especially cogent regarding the idea of 'progress'. Alternatively, there are numerous other books specifically about the development of the idea of 'progress'—a word with almost as many conflicting meanings as 'creativity'.
    - Dawkins, Richard. *The Selfish Gene* – This is the book that brought the implications of evolutionary theory home. The harsh truths of how nature works so clearly explicated are redeemed by Dawkins' virtually child-like wonder, awe, and appreciation of the world we live in.
    - Dennett, Daniel C. *Darwin's Dangerous Idea: Evolution and the Meanings of Life* – The number of books on evolutionary theory almost rival the number of books purporting to disprove it. (Many of the former are worth reading, very few of the latter.) Dennett's is the best discussion of evolutionary theory that I've read. It both covers all the scientific ground and evaluates the philosophical implications of Darwin's important insight that natural selection explains so very, very much.

- o God. _Qur'an, Old Testament Bible, New Testament Bible, Bhagavad-Gita, the Eddas, the Sutras, Dianetics, etc._ – I'm not kidding—except about Hubbard's Scientology 'Bible', _Dianetics._ The various sacred texts are worth a close and critical reading—although a cynic might say _just_ to cure one of naïve religiosity, which certainly is one of the usual results. However, the classic sacred texts do contain much wisdom buried amidst the strange and silly and boring bits. There is no question that they are works of great creativity, and in the context of this chapter, they are relevant for revealing the evolution of very important ideas—including morality, creativity, and purpose.
  - o Porter, Lewis. _John Coltrane: His Life and Music_ – Serious musical scholarship and biography of 'classical' music and its creators is easy to find. Unfortunately, the books on the creative lives and works of individuals working in the jazz genre don't as often reach the same standards of excellence. This book does.
  - o Wilson, E. O. _Consilience: The Unity of Knowledge_ and _The Creation: An Appeal to Save Life on Earth_ – _Consilience_ is a clarion call for the unification of the diverse disciplines of science and even art. _The Creation_ is a desperate appeal to the religious community to join with the scientific community, despite their philosophical differences, to save our planet. It is ironic and very sad that Wilson, who is the most conciliatory and most tactfully gentle of scientific thinkers, has been so demonized by rabid ideologists, both from the religious right and the sanctimonious left.

- **Math, The Old Path, Newly Discovered**
  - General suggestions and remarks.
    - o In positioning _Secret Agents_ as a discourse about the relationship of creative art and of creative science, mathematics may seem to be left out at the periphery—a place it seems to occupy in most people's minds. While pure math isn't science, isn't empirical, is in fact closer to philosophy or abstract music than anything else (as most mathematicians will tell you), in the context of my argument, I must clump it with science; for it like science, lives on the

opposite side of the two culture divide from the arts. And ironically it once lived, perhaps even more than science, deep in the heart of all the arts, especially the visual and musical. My recommendations here are to works that capture some of the aesthetic magic of mathematics and do something to reinstate its status as something other than mere arithmetic or an abstruse, unintelligible academic subject.

- Recommended reading.
  o Bool, F. H. Kist, J.R., and Wierda F. *M.C. Escher: His Life and Complete Graphic Work* – Almost any of the collections of Escher's artworks is worth having, but everyone seems to agree that this is the biggest and best—and most comprehensive, for it includes detailed biographical and interpretative information as well as the images of his work.
  o Gleick, James. *Chaos* – Intimately related to fractal theory is chaos theory, and once again Gleick demonstrates why he is one of our premier science writers. In this book he clearly explains in layman's terms the fascinating ideas and people behind the new science of "chaos". It was this wonderful book that drove me out of my bathtub (where I was reading it) to hack in a simple BASIC program into my computer that demonstrated one of the fundamental principles of this field.
  o Hofstadter, Douglas. *Godel, Escher, Bach: An Eternal Golden Braid* – This modern classic is required reading for anyone interested in the complex relationship of art, science, philosophy, and mathematics. Hofstadter is a modern Renaissance Man and a philosopher of the first order. He also is a fine writer, capable of making even the most abstruse and mind-twisting ideas accessible.
  o Mandelbrot, Benoît. *The Fractal Geometry of Nature* – This is Mandelbrot's magnum opus and the book that put fractals on the multi-dimensional map—so to speak. This very influential book may no longer be the best source of detailed information about fractals, but it remains worthy of attention.
  o Nasar, Sylvia. *A Beautiful Mind* – The well known movie was excellent but so is the book. I could have just as easily recommended this story of the brilliant

mathematician John Nash who suffered from schizophrenia in reference to the chapter on madness and genius. No matter, for it is a book highly relevant to the themes of *Agents* and what matters is that I *do* include it.

- o Paulos, John Allen. *Innumeracy: Mathematical Illiteracy and Its Consequences* – This very readable book (even for those who may almost be innumerate) should convince anyone that math is as central to a full apprehension of the world as is literacy. In another of his books (*Once Upon a Number : The Hidden Mathematical Logic of Stories*) he explores the relationship between math and art, concluding it with the same call for resolution of the two cultures conflict that I have been making in *Secret Agents*.

- **The Shock Of Getting Wired**
  - General suggestions and remarks.
    - o There now is so much good writing about computers and their influence that it is difficult to make any kind of modest selection. Furthermore, much of it is so current as not to be available in book format. Strongly recommend for up-to-date— and both scientifically and artistically savvy—reading is the periodical *Wired*.
  - Recommended reading.
    - o Davis, Martin. *The Universal Computer: The Road from Leibniz to Turing* – This is a very readable history of the development of the major ideas relating to the computer revolution, as well as a rich source of interesting information about the creative individuals behind these ideas.
    - o McLuhan, Marshall and Fiore, Quentin. *The Medium is the Massage* – this book, along with the earlier *Understanding Media*, focuses on the effects of communication media independent of their content. Opinions about McLuhan tend to be polarized: people tend to either consider McLuhan a genius with profound insights into how the new popular media are affecting us or as an over-blown pseudo-intellectual academic that didn't know what he was talking about. Suffice it to say that any creative person's work that inspires such lavish praise or dismissive opprobrium is worth attention.

- Postman, Neil. *Technopoly: The Surrender of Culture to Technology* – There isn't very much in this diatribe against technology with which I would agree, but Postman is among the most articulate of contemporary Luddites. The issues he raises are ones that deserve consideration and evaluation. In many cases reasoned rebuttal is required, for Postman cannot be dismissed as a reactionary flake: his concerns are rooted in a profound and deeply felt humanism.
- Wozniak, Steve and Smith, Gina. *iWoz: From Computer Geek to Cult Icon: How I Invented the Personal Computer, Co-Founded Apple, and Had Fun Doing It* – This book is by Jobs' less-than-humble partner, "The Woz", about the initial development of the Apple home computer. It is more relevant to the theme of creativity in science than the various books about Jobs' himself, which tend to focus on his entrepreneurial skills and the ongoing competition between Microsoft and Apple.

- **Monkeys At Their Word Processors**
  - General suggestions and remarks.
    - The developments in artificial intelligence have failed to live up to early optimistic predictions, but they are nonetheless substantial and impressive. What is of particular interest is the work being done to make computers 'artificially creative'. I've included in this listing some of the major early philosophical arguments regarding this issue. Only time will tell who was right.

  - Recommended reading.
    - Brighton, Henry and Selina, Howard. *Introducing Artificial Intelligence* – This is a recently published entertaining, irreverent, and informative overview of the whole field of artificial intelligence.
    - Minsky, Marvin. *Society of Mind* – This is undoubtedly one of the most famous and controversial attempts to understand 'mind', and the potential for non-humans to have one. It is an entertaining look into one of the more brilliant and idiosyncratic minds that has ever dealt with this issue..
    - Searle, J. R. *Mind: A Brief Introduction* – Searle, of Chinese Room fame, is dubious about the claims of

the so-called 'strong' AI-ers, and this fairly recent book presents the philosophical bases and biases that underpin his view of what intelligence and human consciousness is all about.

o Shermer, Michael. *Why Darwin Matters: The Case Against Intelligent Design* – I include this book here because God was my final case study, and natural selection is the ultimate form of 'artificial' intelligence.

o Turing, Alan (Copeland, ed.) *The Essential Turing: Seminal Writings in Computing, Logic, Philosophy, Artificial Intelligence, and Artificial Life plus The Secrets of Enigma* – This is not a book the casual reader will want to read cover to cover, but for those interested in the thoughts of this genius who if justifiably considered the father of modern computing and artificial intelligence, it is certainly worth dipping into.

o Weizenbaum, Joseph. *Computer Power and Human Reason: From Judgment To Calculation* – This book by MIT Professor Emeritus of Computer Science is a philosophical examination of the implications of advances in artificial intelligence. Weizenbaum is the author of the famous computer program *Eliza*, mentioned in this chapter, which simulates a client-centred psychotherapist. (Freeware versions of this relatively simple program, with source code, can be found on the Internet.)

AUTHOR'S NOTE

"I believe that literature, like science, is a way of exploring different perspectives; and I believe that the results of these literary explorations, like the results of science, are always inherently tentative. It is for this reason that I choose to call my major works hypotheses. *Secret Agents Future: Going Where There Be Dragons* is Hypothesis 18."

ABOUT THE AUTHOR

Ken Stange is the author of 17 books of poetry, fiction, and non-fiction, as well as hundreds of publications in literary and scientific journals. He was the winner of the 2011 Exile/Vanderbilt prize for short fiction, and he is also a visual artist and Professor Emeritus at Nipissing University where he continues to teach "The Psychology of Art" as an online course. His special interest is the relationship of art and science and creativity.

www.ingramcontent.com/pod-product-compliance
Lightning Source LLC
Chambersburg PA
CBHW070803280326
41934CB00012B/3030